INFRARED
SPECTRA
OF
PESTICIDES

INFRARED SPECTRA OF PESTICIDES

TOM VISSER

*National Institute of Public Health
and Environmental Protection
Bilthoven, The Netherlands*

Marcel Dekker, Inc.   New York • Basel • Hong Kong

Library of Congress Cataloging-in-Publication Data

Visser, Tom
    Infrared spectra of pesticides / Tom Visser
       p.   cm.
    Includes bibliographical references and indexes.
    ISBN 0-8247-8988-1 (alk. paper)
    1. Pesticides--Spectra.   2. Infrared spectra.   I. Title.
    SB960.V57   1993
    668'.65--dc20   92-39856
                                          CIP

This book is printed on acid-free paper.

MARCEL DEKKER, INC.
270 Madison Avenue, New York, New York 10016

Current printing (last digit):
10 9 8 7 6 5 4 3 2 1

PRINTED IN THE UNITED STATES OF AMERICA

Preface

Spectral libraries have proven to be valuable in the structural analysis of unknown analytes, as they offer insight into the absorption characteristics of functional groups and closely related compounds. Moreover, the unique discriminating properties of infrared spectra make appropriate reference collections of this, a powerful tool for confirming the identity of a sample. Many infrared spectral libraries have been published over the years; in this respect, a new collection of infrared spectra of pesticides is nothing special. On the other hand, recent developments have made publication of this particular atlas relevant.

First, the majority of infrared spectrometric analyses are now performed on Fourier transform instruments. As a consequence, the quality of the spectra in terms of signal to noise ratio and optical resolution has improved greatly compared to those recorded on dispersive spectrometers. In addition, data manipulation features have affected the presentation of the spectra, which makes them difficult to compare to reference data recorded on conventional machines.

Second, the continuous demand for more effective and specialized pesticides has led to a series of new products. For many of these compounds and related metabolites, infrared spectral data have never been published.

Finally, the growing importance of good laboratory practices (GLP) and good manufacturing practices (GMP) has greatly increased the application of infrared spectroscopy to control of standards in identification. In the National Institute of Public Health and Environmental Protection of The Netherlands, it has become a standard GLP procedure to confirm the identity of pesticides, prior to performing qualitative and quantitative analysis. Actually, this book is a spin-off of this procedure. It is primarily meant to be used as a quick reference guide, although it may also fulfill demands concerning identification and structural elucidation of pesticides. As such it will be a useful tool on the bench of spectroscopists working in many analytical, environmental, agricultural, pharmaceutical, and toxicological disciplines.

Without the help of many people this book would never have come to its completion. Therefore I would like to gratefully acknowledge Dr. Peter Greve, Huub van de Broek, Ramon Ramlal, Rob Zwartjes and other fellows of the Laboratory for Organic-analytical Chemistry for bringing together so much physical and chemical data. Special thanks are owed to Marjo Vredenbregt for carefully recording many of the spectra and for encouraging me to carry out the job.

                                                        Tom Visser

Contents

v

Introduction

Identification and quantitation of pesticides is subject to stringent forensic and regulatory requirements because of their toxic properties. The analysis is based on the use of standards and for that reason accurate quality control of these compounds is an important part of the analytical procedure. Infrared spectrometry is commonly applied for this purpose due to the discriminating properties of this technique. Assessment of the correctness of the identification is usually based on the similarity between the recorded spectrum and a reference one. For this purpose collections of reference data are indispensable. However, spectral libraries of currently used pesticides and related metabolites are not commercially available. This book was compiled to accommodate this demand.

Samples

All materials were commercially available products. The correctness of their identification was verified by (1) interpretation of the IR spectrum, (2) comparison with a previously published IR spectrum, and (3) control of chemical or physical data.

Sample preparation

Liquid samples were run as a neat film at ambient temperature in a demountable cell between NaCl or KBr windows. The pathlength was variable and set such that the strongest peak in the spectrum reached an absorbance value between 0.5 and 3 absorbance units.

Solid samples were run as 13 millimeter diameter KBr pellets at ambient temperature. Pellets were prepared by carefully grinding, a 1 to 3 milligram sample with subsequent adding and mixing of 300 milligrams dry KBr powder. Similar to the liquid samples, the concentration was aimed to be such that the strongest peak in the spectrum reached an absorbance value between 0.5 and 3 absorbance units. Pellets were pressed under vacuum at 200 kilopascal per square centimeter.

There were two reasons for using the KBr pelleting method instead of the mull technique. First, quality control requires insight into a spectral region as wide as possible. Mull spectra, however, show interfering bands of paraffin. Second, KBr spectra compare much better to spectra recorded by means of diffuse reflectance (DRIFT) and cryotrapping GC-FTIR, two techniques which are likely to be commonly applied in the near future.

Samples with a melting point between 20^0 and 40^0 C were prepared as a melt in a demountable cell between alkali halide windows.

A few pesticides have been analyzed with the relatively new technique of cryotrapping gas chromatography/FTIR. These standards were highly diluted solutions and could not be prepared with any other technique. It is noted that cryotrapped spectra of solids compare very well with KBr spectra .

Instrumentation

Two systems were used for the recording of the spectra. The majority of the data have been recorded on a Bruker IFS-85 Fourier transform instrument equipped with a Bruker Aspect 1000 computer for data processing. The detector was DTGS and the applied optical resolution was 2 cm^{-1} with 32 scans coadded. Some of the older spectra were recorded at 4 cm^{-1} resolution.

A few spectra were measured by means of a Biorad Digilab cryotrapping GC/FTIR system consisting of a Digilab FTS-40 interferometer with a Tracer GC/FTIR interface equipped with a narrow range MCT detector. The interface was connected to a Carlo-Erba 5160 high resolution gas chromatograph. Data processing was performed on a Digilab SPC 3200 computer. Cryotrapped spectra were recorded in post-run measurement, with an optical resolution of 2 cm^{-1}, 512 scans coadded.

Presentation and information

Fair comparison of infrared spectra requires identical presentation. For that reason the spectra in this book have been normalized to a transmission value of 3% (1.5 absorbance units) of the strongest band in the spectrum. Spectra have been plotted in transmittance as it offers better insight into the presence of weak skeletal bands than presentation in absorbance. A table with band maxima and corresponding relative intensity values is included.

Limited chemical and physical information has been added to the spectral pages. The molecular structure, brutoformula, molecular ion weight and the exact mass are indicated. The chemical name has been reported according to the IUPAC nomenclature. If not available, Chemical Abstract names have been used instead. CAS numbers, pesticide type and most common trivial and trade names have been given as far as possible. Information was derived from the pesticide databank of the laboratory for Organic-analytical Chemistry of the National Institute of Public Health and Environmental Protection of The Netherlands. Additional data were extracted from The Agrochemicals Handbook [1].

The order of presentation of the data is primarily alphabetical. Spectra of metabolites appear after the spectrum of the original compound. Alphabetic and brutoformula indexes are included.

Reference

1.   The Agrochemicals Handbook , The Royal Society of Chemistry, University of Nottingham, England 1986

Infrared Spectral Data

COMPOUND : **Acephate**

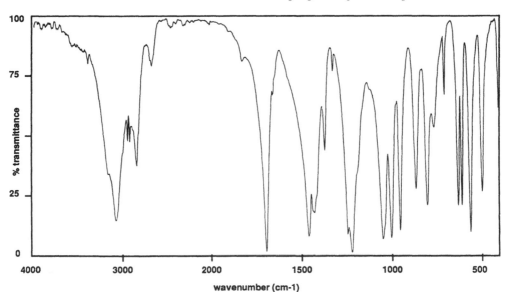

Chemical name   : O,S-dimethyl acetylphosphoramidothioate

Other name	: Orthene, RE 12420, Ortho 12420		
Type	: insecticide		
Brutoformula	: C4H10NO3PS	CAS nr	: 30560-19-1
Molecular mass	: 183.16296	Exact mass	: 183.0118982
Instrument	: Bruker IFS-85	Optical resolution	: 2 cm-1
Scans	: 32	Sampling technique	: KBr pellet

Band maxima with relative intensity :

492	74	560	91	605	81	624	81	701	44
760	47	798	80	863	73	954	90	1002	94
1049	94	1223	100	1326	25	1373	57	1433	83
1462	92	1699	99	2682	20	2854	62	2933	53
2956	52	3089	85						

COMPOUND : **Aclonifen**

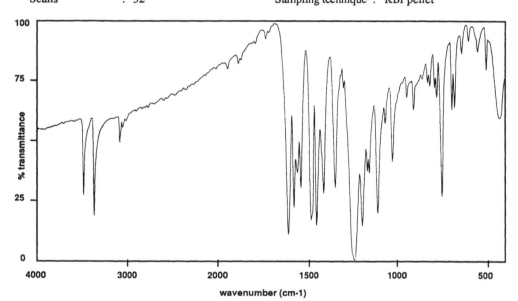

Chemical name : 2-chloro-6-nitro-3-phenoxyaniline

Other name : Bandur, CME 127, Bander
Type : herbicide
Brutoformula : C12H9ClN2O3   CAS nr : 74070-46-5
Molecular Mass : 264.6701   Exact mass : 264.030163
Instrument : Bruker IFS-85   Optical resolution : 2 cm-1
Scans : 32   Sampling technique : KBr pellet

Band maxima with relative intensity :

438	40	509	21	556	12	606	8	644	13
684	36	700	38	757	73	784	32	795	28
821	27	910	37	946	32	1028	59	1070	43
1114	80	1163	64	1173	63	1203	86	1248	100
1358	69	1423	72	1463	85	1494	83	1551	70
1570	63	1592	78	1623	89	3105	50	3388	81
3502	72								

COMPOUND : **Alachlor**

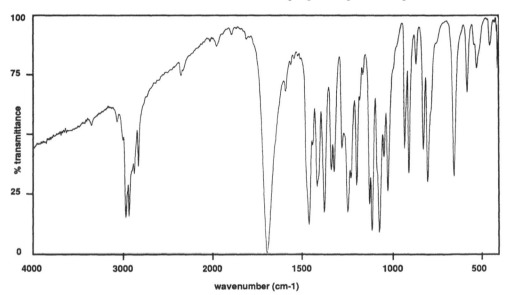

Chemical name   :   2-chloro-2',6'-diethyl-N-(methoxymethyl)acetanilide

Other name   :   Metachlor, CP 50144, Lasso
Type   :   herbicide
Brutoformula   :   C14H20ClNO2   CAS nr   :   15972-60-8
Molecular Mass   :   269.774   Exact mass   :   269.1182465
Instrument   :   Bruker IFS-85   Optical resolution   :   2 cm-1
Scans   :   32   Sampling technique :   KBr pellet

Band maxima with relative intensity :

443	12	516	22	578	32	644	68	792	70
817	56	862	20	903	67	927	57	1017	74
1039	59	1064	91	1105	90	1119	80	1192	72
1241	83	1273	56	1317	66	1332	65	1370	83
1409	72	1453	88	1688	100	2830	63	2934	84
2969	85								

3

COMPOUND : **Aldicarb**

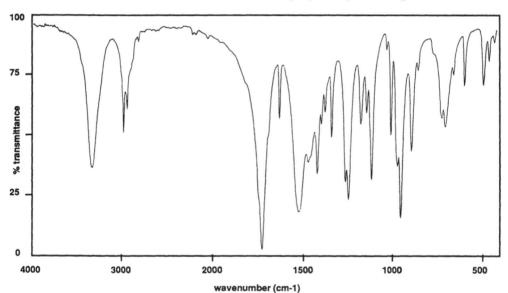

Chemical name	:	2-methyl-2-(methylthio)propionaldehyde O-methylcarbamoyloxime		
Other name	:	Temik, UC 21149		
Type	:	insecticide, acaricide, nematicide		
Brutoformula	:	C7H14N2O2S	CAS nr	: 116-06-3
Molecular Mass	:	190.26183	Exact mass	: 190.0775917
Instrument	:	Bruker IFS-85	Optical resolution	: 2 cm-1
Scans	:	32	Sampling technique	: KBr pellet

Band maxima with transmittance value :

453	22	484	31	588	32	693	48	881	58
938	86	995	53	1104	71	1133	43	1165	47
1234	79	1251	72	1332	54	1369	43	1411	68
1513	84	1624	46	1717	100	2926	42	2965	52
3312	66								

4

COMPOUND : **Aldicarb-sulfon**

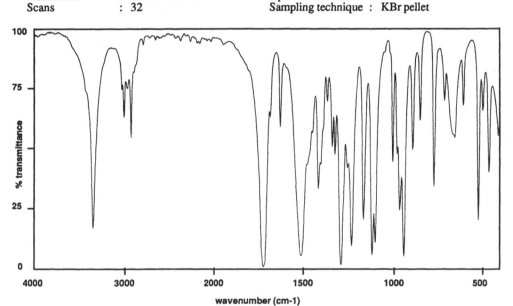

Chemical name	: 2-methyl-2-(methylsulphonyl)-propanal-O-(methylamino) carbonyloxim

Other names   : -
Type   : metabolite of Aldicarb
Brutoformula   : C7H14N2O4S   CAS nr   : -
Molecular Mass   : 222.2606   Exact mass   : 222.067419
Instrument   : Bruker IFS-85   Optical resolution   : 2 cm-1
Scans   : 32   Sampling technique   : KBr pellet

Band maxima with relative intensity :

458	60	491	34	517	80	602	32	649	45
706	29	767	66	845	38	889	50	941	95
964	76	977	52	1001	56	1103	89	1120	94
1164	80	1234	91	1292	98	1325	53	1341	49
1368	30	1416	67	1511	95	1629	41	1724	100
2944	45	3019	36	3356	84				

5

COMPOUND : **Aldicarb-sulfoxide**

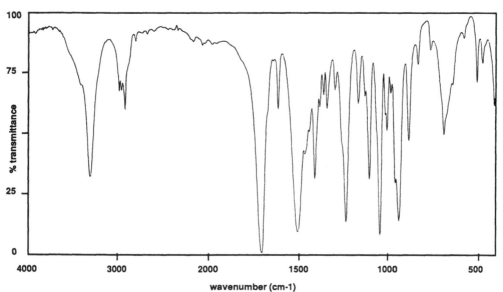

Chemical name   : 2-methyl-2-(methylsulphoxyl)-propanal-O-(methylamino) carbonyloxim

Other names   : -
Type   : metabolite of Aldicarb
Brutoformula   : C7H14N2O3S   CAS nr   : -
Molecular Mass : 206.2612   Exact mass   : 206.0725058
Instrument   : Bruker IFS-85   Optical resolution   : 2 cm-1
Scans   : 32   Sampling technique : KBr pellet

Band maxima with relative intensity :

476	20	507	28	692	50	835	20	886	52
941	86	1003	49	1047	92	1106	69	1167	37
1238	87	1298	31	1344	39	1363	33	1414	69
1514	91	1623	39	1716	100	2929	39	3328	67

COMPOUND : **Aldrin**

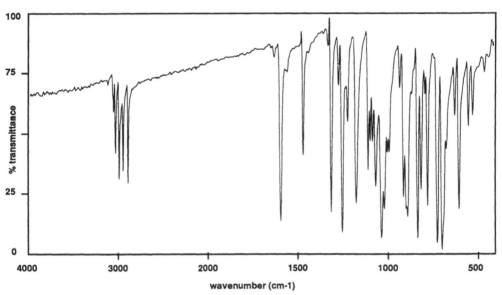

Chemical name : (1R,4S,5S,8R)-1,2,3,4,10,10-hexachloro-1,4,4a,5,8,8a-hexahydro-
1,4 : 5,8-dimethanonaphtalene
Other name : HHDN
Type : insecticide
Brutoformula : C12H8Cl6   CAS nr   : 309-00-2
Molecular Mass : 364.91556   Exact mass   : 361.8757154
Instrument : Bruker IFS-85   Optical resolution : 2 cm-1
Scans : 32   Sampling technique : KBr pellet

Band maxima with relative intensity :

459	27	524	46	548	50	598	83	622	46
694	100	720	96	773	82	786	37	811	76
831	94	888	87	911	79	932	34	990	61
1001	61	1017	84	1032	95	1065	75	1082	56
1096	56	1108	68	1175	81	1222	48	1251	92
1272	33	1313	84	1330	17	1469	62	1597	88
1635	21	2900	73	2955	68	2999	71	3039	61
3060	44								

COMPOUND : **Alloxydim-sodium**

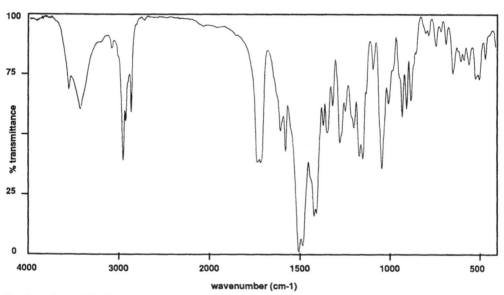

Chemical name	: methyl 3-[1-(allyloxyimino)butyl]-4-hydroxy-6,6-dimethyl- 2-oxocyclohex-3-enecarboxylate, sodium salt

Other names	: -
Type	: herbicide
Brutoformula	: C17H24NO5Na
Molecular Mass	: 345.3743
Instrument	: Bruker IFS-85
Scans	: 32

CAS nr	: 55635-13-7
Exact mass	: 345.1552151
Optical resolution	: 2 cm-1
Sampling technique	: KBr pellet

Band maxima with relative intensity :

406	14	465	18	498	26	556	20	602	18
645	24	683	11	739	12	780	8	879	35
903	39	927	43	1004	37	1042	64	1092	22
1151	60	1171	59	1200	47	1279	53	1319	38
1352	49	1371	46	1422	84	1506	100	1580	57
1608	48	1720	62	2870	41	2958	61	3432	39
3557	31								

8

COMPOUND : **2-amino-N-isopropylbenzamide**

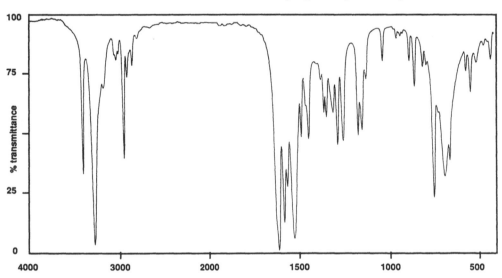

Chemical name   : 2-amino-N-isopropylbenzamide

Other names   : -
Type   : metabolite of bentazone
Brutoformula   : C10H14N2O   CAS nr   : 30391-89-0
Molecular Mass   : 178.2358   Exact mass   : 178.110604
Instrument   : Bruker IFS-85   Optical resolution   : 2 cm-1
Scans   : 32   Sampling technique   : KBr pellet

Band maxima with relative intensity :

429	20	541	34	566	25	660	63	688	69
749	78	808	22	857	31	887	20	1036	20
1154	49	1175	52	1257	53	1289	55	1313	42
1350	44	1364	42	1449	53	1489	52	1533	95
1569	73	1588	88	1619	100	2872	20	2930	25
2970	60	3050	18	3300	96	3417	66		

9

COMPOUND : **2-Amino-benzimidazole**

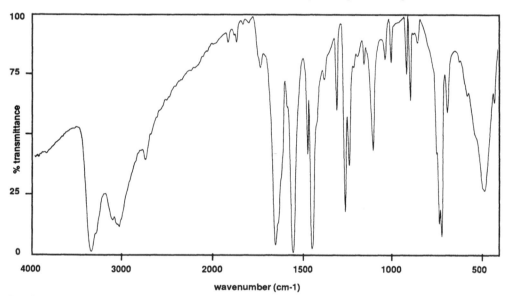

Chemical name   : 2-amino-benzimidazole

Other names	: -		
Type	: metabolite of Carbendazim		
Brutoformula	: C7H7N3	CAS nr	: 934-32-7
Molecular Mass	: 133.1539	Exact mass	: 133.063991
Instrument	: Bruker IFS-85	Optical resolution	: 2 cm-1
Scans	: 32	Sampling technique	: KBr pellet

Band maxima with relative intensity :

492	74	693	41	729	93	742	88	897	36
918	25	1001	20	1034	18	1109	57	1157	21
1244	63	1269	82	1313	40	1460	98	1481	58
1568	100	1667	96	3054	88	3381	98		

COMPOUND : **Amino-methylphosphonic   acid**

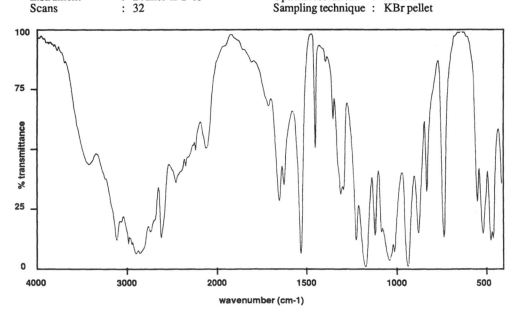

Chemical name   : amino-methylphosphonic acid

Other names   : -
Type   : metabolite of Glyphosate
Brutoformula   : CH6NO3P   CAS nr   : 1066-51-9
Molecular Mass   : 111.0376   Exact mass   : 111.0085271
Instrument   : Bruker IFS-85   Optical resolution : 2 cm-1
Scans   : 32   Sampling technique : KBr pellet

Band maxima with relative intensity :

466	89	510	86	540	72	728	87	821	68
871	85	931	100	1034	97	1112	86	1166	99
1216	88	1300	69	1342	37	1442	50	1528	94
1622	65	1650	71	2116	49	2455	63	2618	86
2899	93	3112	87						

11

COMPOUND : **Amitraz**

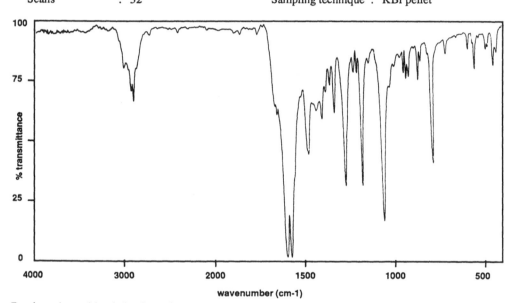

Chemical name	:	N-methyl-bis(2,4-xylyliminomethyl)-amine

Other name : Triazid, Tetranyx, BTS 27419, RD 27419
Type : acaricide, insecticide
Brutoformula : C19H23N3   CAS nr   : 33089-61-1
Molecular Mass : 293.41526   Exact mass   : 293.1891854
Instrument : Bruker IFS-85   Optical resolution : 2 cm-1
Scans : 32   Sampling technique : KBr pellet

Band maxima with relative intensity :

461	20	501	13	565	21	600	13	724	15
805	61	881	25	946	25	961	22	1081	85
1201	70	1222	22	1242	22	1294	70	1351	39
1374	27	1419	41	1494	56	1598	100	1622	99
2917	32								

COMPOUND : **Amitrole**

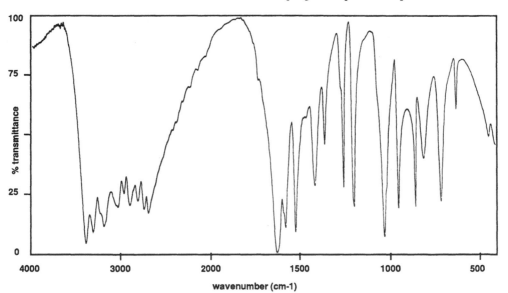

Chemical name   : 5-amino-1,2,4-triazole

Other name   : Aminotriazole
Type   : herbicide
Brutoformula   : C2H4N4   CAS nr   : 61-82-5
Molecular Mass   : 84.08098   Exact mass   : 84.0435912
Instrument   : Bruker IFS-85   Optical resolution : 2 cm-1
Scans   : 32   Sampling technique : KBr pellet

Band maxima with relative intensity :

405	59	461	52	642	40	730	79	829	60
877	81	968	81	1046	93	1213	80	1270	72
1372	54	1427	71	1536	91	1594	89	1642	100
2724	82	2772	80	2840	77	2930	79	2990	74
3057	79	3215	87	3332	89	3412	94		

COMPOUND : **Asulam**

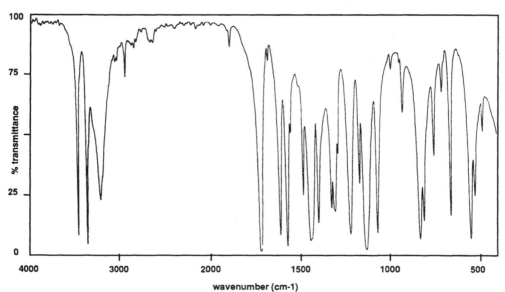

Chemical name : methyl 4-aminophenylsulphonylcarbamate

Other names	: MB 9057		
Type	: herbicide		
Brutoformula	: C8H10N2O4S	CAS nr	: 3337-71-1
Molecular Mass	: 230.2399	Exact mass	: 230.0361215
Instrument	: Bruker IFS-85	Optical resolution	: 2 cm-1
Scans	: 32	Sampling technique	: KBr pellet

Band maxima with relative intensity :

504	50	547	77	572	95	681	85	728	33
774	60	832	87	854	95	946	41	1006	23
1090	92	1151	99	1188	71	1241	93	1325	83
1346	82	1419	88	1463	95	1503	76	1594	97
1632	93	1744	100	1906	13	2649	10	2955	25
3245	77	3389	95	3487	92				

14

COMPOUND : **Atrazin**

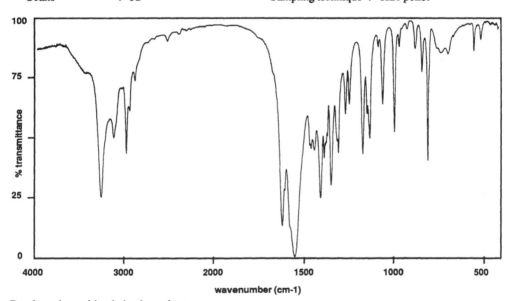

Chemical name   : 2-chloro-4-ethylamino-6-isopropylamino-1,3,5-triazine

Other names   : Primatol
Type   : herbicide
Brutoformula   : C8H14ClN5   CAS nr   : 1912-24-9
Molecular Mass : 215.68728   Exact mass   : 215.0937635
Instrument   : Bruker IFS-85   Optical resolution : 2 cm-1
Scans   : 32   Sampling technique : KBr pellet

Band maxima with relative intensity :

514	8	553	12	696	14	806	59	838	21
878	11	992	47	1056	35	1128	49	1167	56
1244	35	1265	39	1304	56	1346	69	1383	58
1404	75	1437	54	1552	100	1621	86	2973	55
3115	49	3257	74						

15

COMPOUND : **Azamethiphos**

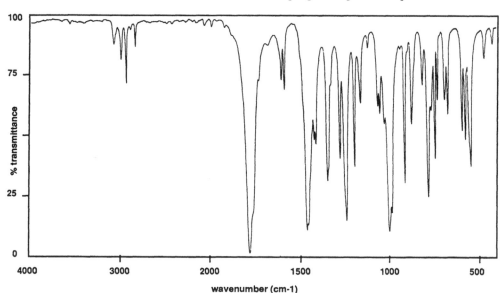

Chemical name	: O,O-dimethyl S-[(6-chloro-2-oxo-oxazolo[4,5-b]pyridin-3(2H)-yl)methyl] phosphorothioate
Other name	: Azamethifos, CGA 18809
Type	: insecticide

Brutoformula	: C9H10ClN2O5PS	CAS nr	: 35575-96-3
Molecular Mass	: 324.67721	Exact mass	: 323.9736527
Instrument	: Bruker IFS-85	Optical resolution	: 2 cm-1
Scans	: 32	Sampling technique	: KBr pellet

Band maxima with relative intensity :

438	12	484	18	561	64	590	52	607	49
685	42	703	36	742	36	757	60	797	77
826	29	888	46	927	71	1016	91	1069	41
1081	38	1136	13	1179	37	1214	63	1262	86
1296	60	1366	69	1428	53	1483	90	1601	30
1618	26	1799	100	2851	12	2952	27	3011	17
3090	10								

16

COMPOUND : **Azinphos-ethyl**

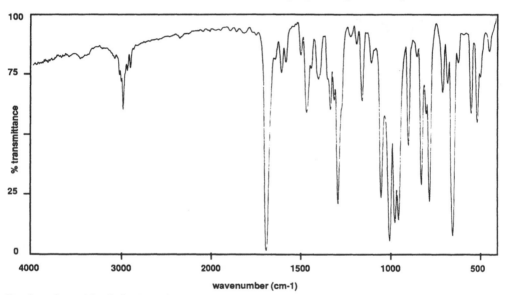

Chemical name	:	S-(3,4-dihydro-4-oxobenzo[d]-[1,2,3]-triazin-3-ylmethyl) O,O-diethyl phosphorodithioate		
Other name	:	Ethylguthion, Ethylgusathion		
Type	:	acaricide, insecticide		
Brutoformula	:	C12H16N3O3PS2	CAS nr	: 2642-71-9
Molecular mass	:	345.37338	Exact mass	: 345.0370649
Instrument	:	Bruker IFS-85	Optical resolution	: 2 cm-1
Scans	:	32	Sampling technique	: KBr pellet

Band maxima with relative intensity :

514	45	547	41	649	93	680	29	707	32
779	79	825	72	898	55	950	87	972	88
1001	96	1049	77	1105	20	1155	36	1186	12
1286	80	1330	40	1396	27	1463	41	1496	16
1581	19	1606	24	1689	100	2900	22	2979	39

COMPOUND : **Azinphos-methyl**

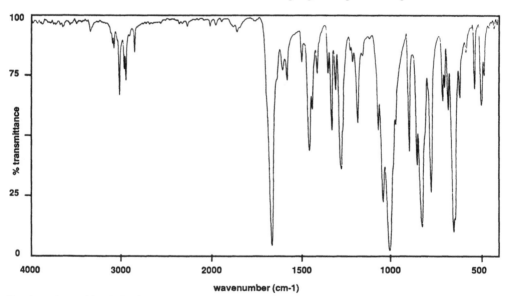

Chemical name	: S-(3,4-dihydro-4-oxobenzo[d]-[1,2,3]-triazin-3-ylmethyl) O,O-dimethyl phosphorodithioate		
Other names	: Guthion, Gusathion		
Type	: acaricide, insecticide		
Brutoformula	: C10H12N3O3PS2	CAS nr	: 86-50-0
Molecular mass	: 317.3192	Exact mass	: 317.0057665
Instrument	: Bruker IFS-85	Optical resolution	: 2 cm-1
Scans	: 32	Sampling technique	: KBr pellet

Band maxima with relative intensity :

500	38	536	32	617	35	659	92	681	40
700	31	710	36	783	75	834	90	857	63
898	58	1009	100	1043	79	1066	48	1184	45
1280	65	1307	31	1331	48	1349	24	1409	24
1459	57	1495	19	1579	27	1677	97	2841	14
2941	26	2958	21	3013	32	3073	12		

COMPOUND : **Azocyclotin**

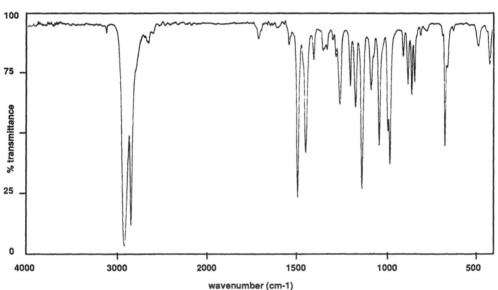

Chemical name   : 1-(tricyclohexylstannyl)-1H-1,2,4-triazole

Other names	: -		
Type	: acaricide		
Brutoformula	: C20H35N3Sn	CAS nr	: 41083-11-8
Molecular mass	: 436.21205	Exact mass	: 437.1852806
Instrument	: Bruker IFS-85	Optical resolution	: 2 cm-1
Scans	: 32	Sampling technique	: KBr pellet

Band maxima with relative intensity :

422	24	483	16	670	58	840	31	859	36
881	32	907	20	982	66	1043	58	1088	34
1138	76	1173	41	1201	32	1258	40	1351	17
1401	21	1445	61	1490	80	1537	15	1706	12
2845	91	2920	100						

COMPOUND : **Barban**

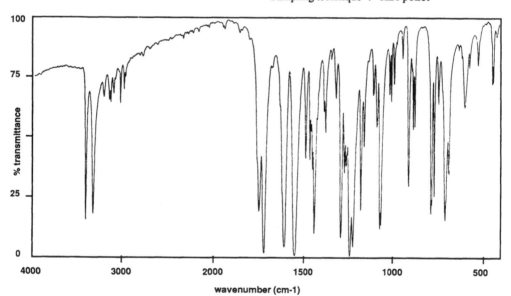

Chemical name	: 4-chlorobut-2-ynyl 3-chlorocarbanilate		

Other name	: Carbyne		
Type	: herbicide		
Brutoformula	: C11H9Cl2NO2	CAS nr	: 101-27-9
Molecular mass	: 258.10588	Exact mass	: 257.001029
Instrument	: Bruker IFS-85	Optical resolution	: 2 cm-1
Scans	: 32	Sampling technique	: KBr pellet

Band maxima with relative intensity :

442	26	594	36	678	64	701	84	737	35
761	74	779	81	867	44	877	46	901	70
981	26	996	34	1005	31	1061	88	1080	45
1100	31	1150	53	1167	80	1212	96	1229	100
1250	61	1259	64	1280	92	1308	33	1366	47
1429	90	1439	63	1455	59	1478	58	1540	99
1599	95	1711	98	1740	80	2958	29	3002	35
3077	31	3111	34	3188	32	3307	82	3388	84

20

COMPOUND : **BAS-43600F**

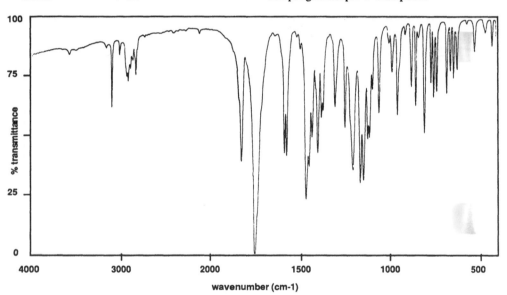

Chemical name   : 3-(3,5-dichlorophenyl)-5-methoxymethyl-5-methyl-2,4-oxazolidindion

Other name   : Meclozoline, Myclozoline, BAS-43603F
Type   : fungicide
Brutoformula   : C12H11NO4Cl2   CAS nr   :
Molecular Mass   : 304.13177   Exact mass   : 303.0065064
Instrument   : Bruker IFS-85   Optical resolution : 2 cm-1
Scans   : 32   Sampling technique : KBr pellet

Band maxima with relative intensity :

430	11	466	5	526	13	622	21	642	24
659	21	679	31	733	30	750	33	764	26
802	48	851	36	875	28	950	40	980	22
1055	39	1092	29	1117	50	1140	68	1159	69
1199	64	1244	46	1301	37	1378	41	1398	56
1430	49	1461	76	1573	57	1586	56	1747	100
1823	60	2820	23	2909	25	3008	14	3095	37

COMPOUND : **Benazolin**

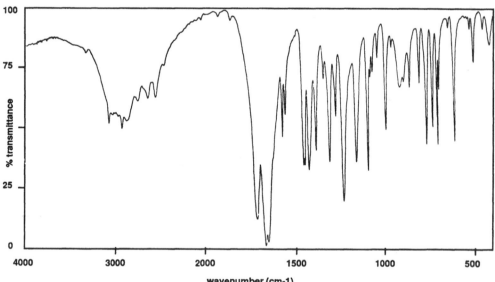

Chemical name : 4-chloro-2-oxobenzothiazolin-3-yl acetic acid

Other names : RD 7693, SN 5482
Type : herbicide
Brutoformula : C9H6ClNO3S
Molecular mass : 243.66607
Instrument : Bruker IFS-85
Scans : 32

CAS nr : 3813-05-6
Exact mass : 242.9756889
Optical resolution : 2 cm-1
Sampling technique : KBr pellet

Band maxima with relative intensity :

508	23	617	57	701	34	710	58	737	50
772	58	810	32	865	33	919	33	999	51
1045	21	1076	27	1106	68	1173	65	1247	81
1287	45	1324	64	1353	29	1399	59	1439	68
1469	66	1566	44	1583	53	1682	100	1729	88
2565	36	2650	36	2940	49				

COMPOUND : **Bendiocarb**

Chemical name	:	2,3-isopropylidenedioxyphenyl N-methylcarbamate

Other names	:	Bendioxocarb, NC 6897			
Type	:	insecticide			
Brutoformula	:	C11H13NO4	CAS nr	:	22781-23-3
Molecular mass	:	223.23056	Exact mass	:	223.0844494
Instrument	:	Bruker IFS-85	Optical resolution	:	2 cm-1
Scans	:	32	Sampling technique	:	KBr pellet

Band maxima with relative intensity :

511	18	571	12	657	26	709	25	733	44
749	23	784	58	795	42	835	39	856	17
875	13	930	47	976	39	1028	63	1046	51
1120	84	1158	69	1219	69	1240	90	1264	88
1362	29	1376	54	1388	38	1419	41	1469	83
1490	80	1536	61	1629	24	1689	33	1720	100
1749	72	2989	18	3332	66				

COMPOUND : **Benodanil**

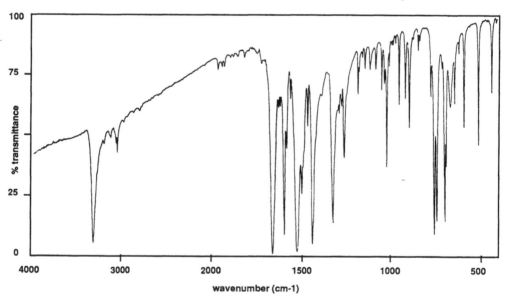

Chemical name   : 2-iodobenzanilide

Other name   : BAS 317F
Type   : fungicide
Brutoformula : C13H10INO   CAS nr   : 15310-01-7
Molecular mass : 323.13525   Exact mass   : 322.9808933
Instrument   : Bruker IFS-85   Optical resolution : 2 cm-1
Scans   : 32   Sampling technique : KBr pellet

Band maxima with relative intensity :

438	32	510	54	588	46	639	36	663	38
694	86	740	86	755	91	774	33	841	14
891	46	912	34	946	36	1014	63	1027	27
1043	30	1078	22	1110	22	1138	21	1178	32
1257	59	1271	37	1322	86	1440	95	1464	45
1499	74	1526	99	1561	34	1582	55	1597	91
1662	100	3037	57	3312	95				

COMPOUND : **Benomyl**

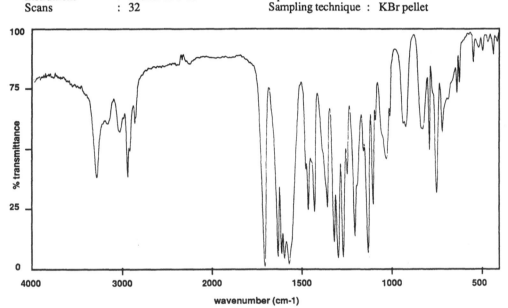

Chemical name : methyl 1-(butylcarbamoyl)benzimidazol-2-ylcarbamate

Other name	: Benlate, INT-1991
Type	: acaricide, fungicide
Brutoformula	: C14H18N4O3
Molecular mass	: 290.32456
Instrument	: Bruker IFS-85
Scans	: 32

CAS nr	: 17804-35-2
Exact mass	: 290.1378779
Optical resolution	: 2 cm-1
Sampling technique	: KBr pellet

Band maxima with relative intensity :

438	11	497	9	548	14	625	22	640	27
724	43	757	69	795	51	835	42	928	41
1036	55	1110	74	1139	94	1213	87	1252	61
1277	96	1305	96	1327	89	1365	75	1437	77
1472	76	1581	98	1607	96	1623	94	1644	95
1718	100	2873	38	2956	61	3041	42	3298	61

COMPOUND : **Bentazone**

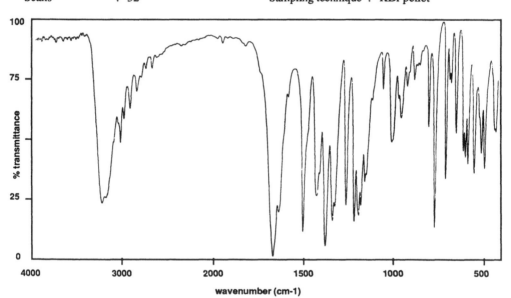

Chemical name   : 3-isopropyl-(1H)-benzo-2 1,3-thiadiazin-4-one 2,2-dioxide

Other name	: Bendioxide, Thianon, Basagran		
Type	: herbicide		
Brutoformula	: C10H12N2O3S	CAS nr	: 25057-89-0
Molecular mass	: 240.27874	Exact mass	: 240.0568566
Instrument	: Bruker IFS-85	Optical resolution	: 2 cm-1
Scans	: 32	Sampling technique	: KBr pellet

Band maxima with relative intensity :

477	61	495	55	534	63	569	60	583	57
595	54	635	47	662	25	691	66	750	87
786	44	867	24	905	27	938	40	987	51
1036	28	1136	67	1158	78	1170	82	1195	84
1241	78	1316	84	1356	95	1406	74	1480	89
1647	100	2806	30	2879	37	2980	52	3181	78

COMPOUND : **Benzoximate**

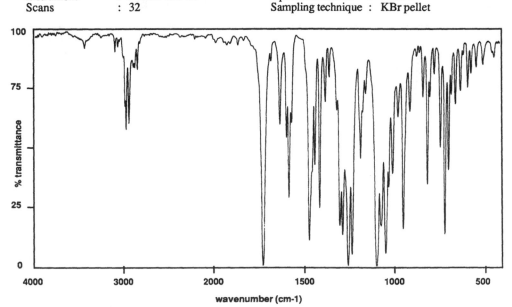

Chemical name : 3-chloro-α-ethoxyimino-2,6-dimethoxybenzyl benzoate

Other name : Benzomate, Citron, Artaban
Type : acaricide
Brutoformula : C18H18ClNO5   CAS nr   : 29104-30-1
Molecular mass : 363.80086   Exact mass   : 363.0873396
Instrument : Bruker IFS-85   Optical resolution : 2 cm-1
Scans : 32   Sampling technique : KBr pellet

Band maxima with relative intensity :

506	14	542	15	572	17	591	23	631	25
658	30	682	26	697	58	717	86	742	49
775	18	799	27	812	65	838	27	909	34
945	84	973	36	1002	60	1024	66	1041	94
1065	82	1090	100	1153	26	1181	54	1227	94
1248	99	1280	86	1297	82	1358	19	1380	30
1411	75	1439	56	1469	89	1586	70	1600	45
1638	39	1729	99	2843	16	2942	39	2974	41

27

COMPOUND : **Benzoylprop-ethyl**

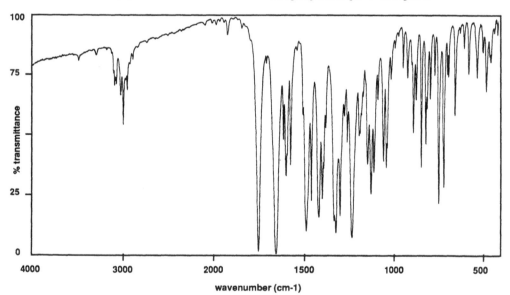

Chemical name   : ethyl N-benzoyl-N-(3,4-dichlorophenyl)-DL-alaninate

Other name   : Suffix, Endaven, WL 17731
Type   : herbicide
Brutoformula   : C18H17Cl2NO3   CAS nr   : 22212-55-1
Molecular mass   : 366.24709   Exact mass   : 365.0585399
Instrument   : Bruker IFS-85   Optical resolution : 2 cm-1
Scans   : 32   Sampling technique : KBr pellet

Band maxima with relative intensity :

467	31	518	25	641	41	676	25	682	24
703	71	731	78	755	24	779	34	804	53
830	63	859	34	874	48	907	25	1001	26
1028	63	1044	60	1076	34	1099	65	1115	74
1135	62	1181	50	1221	93	1247	52	1288	83
1310	91	1368	46	1386	76	1405	84	1446	77
1473	90	1561	62	1585	66	1600	51	1640	100
1736	98	2929	30	2973	45	3003	32	3071	28

28

COMPOUND : **Benzthiazuron**

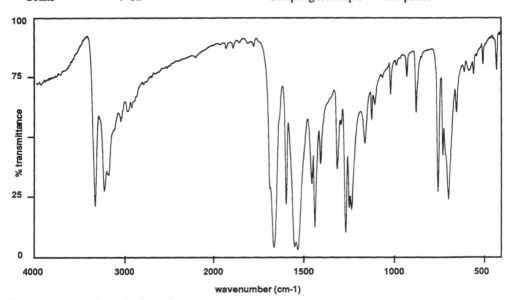

Chemical name   : 1-(benzothiazol-2-yl)-3-methylurea

Other names   : Gatnon, Bayer 60618, Sch 22012
Type   : herbicide
Brutoformula   : C9H9N3OS
Molecular mass   : 207.25158
Instrument   : Bruker IFS-85
Scans   : 32

CAS nr   : 1929-88-0
Exact mass   : 207.0466278
Optical resolution   : 2 cm-1
Sampling technique   : KBr pellet

Band maxima with relative intensity :

402	12	431	23	508	21	560	25	654	41
699	78	730	59	758	75	880	41	932	26
1022	33	1127	44	1164	54	1239	82	1251	82
1271	92	1319	65	1412	63	1445	90	1460	71
1538	100	1603	80	1672	98	3245	74	3344	81

COMPOUND : **Benzyl-adenine**

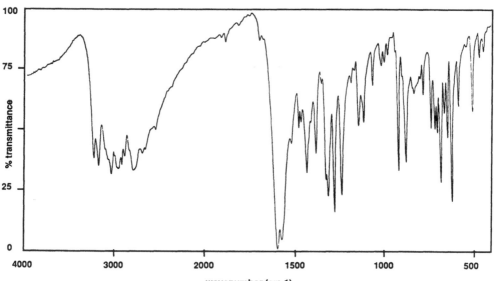

Chemical name : 6-benzyl-adenine

Other names : -
Type : growth regulator
Brutoformula : C12H11N5
Molecular mass : 225.25497
Instrument : Bruker IFS-85
Scans : 32

CAS nr :
Exact mass : 225.1014366
Optical resolution : 2 cm-1
Sampling technique : KBr pellet

Band maxima with relative intensity :

458	17	482	20	522	43	598	40	639	80
660	53	676	43	699	72	716	51	730	50
751	49	793	35	844	35	892	64	933	67
1025	23	1077	31	1129	46	1158	48	1257	77
1299	84	1333	77	1401	59	1453	67	1494	48
1622	100	2821	65	2985	65	3067	67	3207	63
3258	60								

30

COMPOUND : **Beta-HEPO**

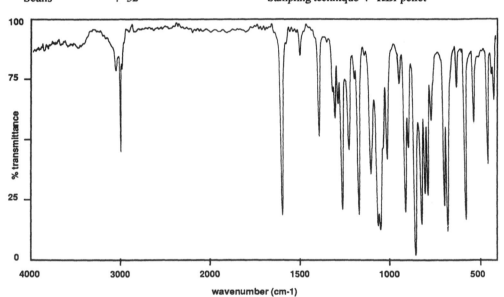

Chemical name : 1,4,5,6,7,8,8-heptachloro-2,3-epoxy-3a,4,7,7a-tetrahydro-4,7-methanoindene

Other name : Heptachloro-epoxide
Type : insecticide
Brutoformula : C10H5Cl7O
Molecular mass : 389.3217
Instrument : Bruker IFS-85
Scans : 32

CAS nr : 102-457-3
Exact mass : 385.816008
Optical resolution : 2 cm-1
Sampling technique : KBr pellet

Band maxima with relative intensity :

415	34	454	61	530	43	579	85	623	28
677	89	695	79	764	42	784	74	801	73
821	86	857	100	892	54	911	81	941	27
1008	59	1048	89	1063	87	1101	65	1168	82
1220	55	1259	80	1278	35	1296	41	1387	49
1492	14	1601	82	2995	54	3039	19		

COMPOUND : **Binapacryl**

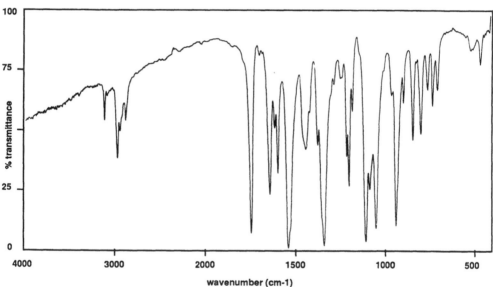

Chemical name : 2-sec.butyl-4,6-dinitrophenyl 3-methylcrotonate

Other name	: Dinoseb-methacrylate, Acricid, Endosan			
Type	: acaricide, fungicide			
Brutoformula	: C15H18N2O6	CAS nr	:	485-31-4
Molecular Mass	: 322.32051	Exact mass	:	322.1164738
Instrument	: Bruker IFS-85	Optical resolution	:	2 cm-1
Scans	: 32	Sampling technique	:	KBr pellet

Band maxima with relative intensity :

467	22	704	33	731	39	759	33	798	51
843	54	896	38	941	90	1055	91	1090	75
1112	97	1186	42	1205	73	1217	61	1252	28
1344	99	1379	56	1444	58	1541	100	1598	68
1616	49	1642	77	1745	93	2878	45	2969	61
3112	45								

COMPOUND : **Bioallethrin**

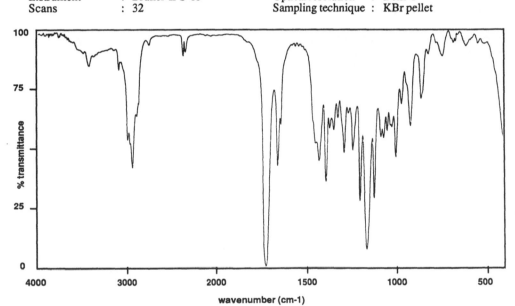

Chemical name   : (+) 3-allyl-2-methyl-4-oxo-cyclopent-2-enyl (+) trans chrysanthemate

Other names   : Trans-allethrin
Type   : insecticide
Brutoformula   : C19H26O3   CAS nr   :
Molecular mass   : 302.41727   Exact mass   : 302.1881819
Instrument   : Bruker IFS-85   Optical resolution   : 2 cm-1
Scans   : 32   Sampling technique : KBr pellet

Band maxima with relative intensity :

741	10	857	28	915	40	964	30	994	53
1044	42	1064	45	1114	71	1154	92	1192	72
1235	50	1282	51	1317	36	1381	64	1421	55
1655	57	1716	100	2918	58	3410	15		

COMPOUND : **Bioresmethrin**

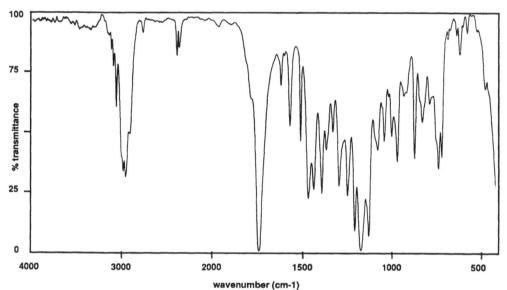

Chemical name   : 2-benzyl (+) trans chrysanthemate

Other names   : Biorestrin, Biobenzylfurine, NRDC 107, RU 11484
Type   : insecticide
Brutoformula   : C22H26O3   CAS nr   : 28434-01-7
Molecular mass   : 338.45072   Exact mass   : 338.1881819
Instrument   : Bruker IFS-85   Optical resolution : 2 cm-1
Scans   : 32   Sampling technique : KBr pellet

Band maxima with relative intensity :

562	7	601	16	700	60	719	65	811	45
855	60	916	34	950	62	983	51	1028	53
1064	57	1114	93	1160	99	1193	91	1235	76
1282	72	1318	49	1353	57	1379	75	1423	73
1453	77	1495	53	1553	47	1603	29	1722	100
2736	7	2923	68	3030	38	3062	21		

COMPOUND : **Biphenox**

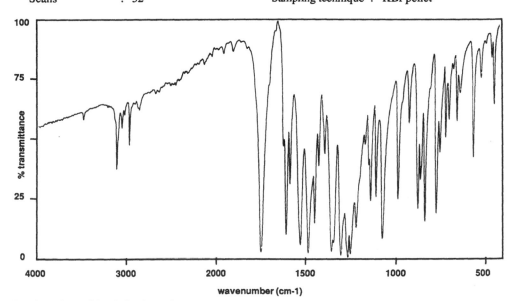

Chemical name : 2-nitro-5-(2,4-dichloro)-phenoxy benzoic acid methyl ester

Other name : Bifenox
Type : herbicide
Brutoformula : C14H9Cl2NO5   CAS nr : 42576-02-3
Molecular mass : 342.13753   Exact mass : 340.9857713
Instrument : Bruker IFS-85   Optical resolution : 2 cm-1
Scans : 32   Sampling technique : KBr pellet

Band maxima with relative intensity :

439	35	453	15	513	23	556	57	631	30
649	42	694	41	713	49	744	55	764	81
828	84	854	67	867	79	917	42	975	75
1062	92	1097	74	1127	76	1206	87	1237	98
1251	100	1290	98	1343	97	1381	56	1415	61
1436	85	1471	98	1515	94	1575	69	1597	90
1740	97	2950	52	3030	45	3086	62		

COMPOUND : **Bitertanol**

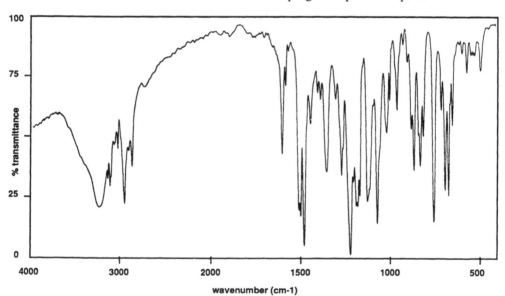

Chemical name	: β-[(1,1'-biphenyl]-4-yloxy)-α-(1,1-dimethylethyl)-1H-1,2,4-triazole-1-ethanol
Other name	: Baycor, Biloaxazol, KWG 0599
Type	: fungicide

Brutoformula	: C20H23N3O2	CAS nr	: 55179-31-2
Molecular mass	: 337.42521	Exact mass	: 337.1790136
Instrument	: Bruker IFS-85	Optical resolution	: 2 cm-1
Scans	: 32	Sampling technique	: KBr pellet

Band maxima with relative intensity :

500	22	532	15	576	23	656	45	677	75
697	72	717	38	761	86	818	49	836	62
870	64	886	46	965	38	1005	34	1023	48
1078	86	1133	77	1175	75	1187	79	1229	100
1276	66	1308	34	1359	64	1392	34	1449	44
1487	96	1506	83	1587	28	1607	57	2871	62
2956	77	3031	54	3120	70	3243	79		

COMPOUND : **Bromacil**

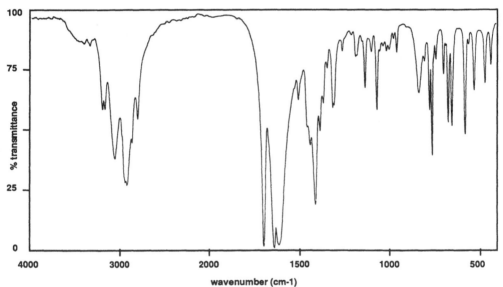

Chemical name : 5-bromo-3-sec.butyl-6-methyluracil

Other names : Hyvar X, Dupont 1976, Uragan
Type : herbicide
Brutoformula : C9H13BrN2O2   CAS nr   : 314040-9
Molecular mass : 261.12016   Exact mass   : 260.0160844
Instrument : Bruker IFS-85   Optical resolution : 2 cm-1
Scans : 32   Sampling technique : KBr pellet

Band maxima with relative intensity :

442	22	476	30	536	33	587	52	659	48
678	47	703	26	745	20	767	61	780	41
839	34	962	17	1074	41	1104	16	1141	32
1192	18	1321	40	1393	50	1420	81	1513	37
1624	98	1651	100	1710	99	2811	44	2939	72
3070	61	3209	40						

COMPOUND : **Bromofenoxim**

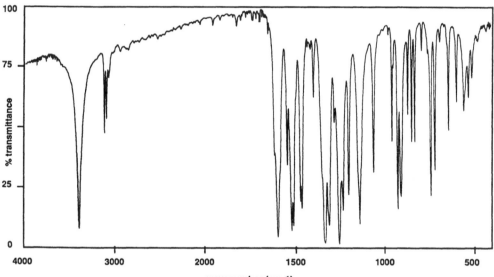

Chemical name	: 3,5-dibromo-4-hydroxybenzaldehyde 2,4-dinitrophenyloxime		

Other name	: Faneron, C 9122		
Type	: hebicide		
Brutoformula	: C13H7O6N3Br2	CAS nr	: 13181-17-4
Molecular mass	: 461.02524	Exact mass	: 458.8702564
Instrument	: Bruker IFS-85	Optical resolution	: 2 cm-1
Scans	: 32	Sampling technique	: KBr pellet

Band maxima with relative intensity :

514	29	534	38	560	43	599	40	644	52
690	14	718	68	741	79	793	18	832	56
849	56	870	45	909	80	927	85	959	56
1067	69	1146	91	1208	79	1240	85	1262	100
1289	48	1319	91	1340	99	1403	37	1469	84
1477	81	1516	92	1526	94	1549	66	1601	96
1822	7	3098	46	3121	52	3400	92		

COMPOUND : **Bromophos**

Chemical name   : O-(4-bromo-2,5-dichlorophenyl) O,O-dimethyl phosphorothioate

Other names   : Omexan, Nexion, OMS 658
Type   : insecticide
Brutoformula   : C8H8BrCl2O3PS   CAS nr   : 2104-96-3
Molecular mass   : 365.99492   Exact mass   : 363.849272
Instrument   : Bruker IFS-85   Optical resolution : 2 cm-1
Scans   : 32   Sampling technique : KBr pellet

Band maxima with relative intensity :

440	45	468	43	526	23	553	32	611	66
632	64	658	42	688	46	717	63	811	93
838	95	883	75	951	87	1029	100	1078	66
1113	52	1182	59	1220	44	1244	58	1342	76
1457	81	1480	33	1553	21	1573	16	1855	15
2846	39	2947	48	3086	44				

COMPOUND : **Bromophos-ethyl**

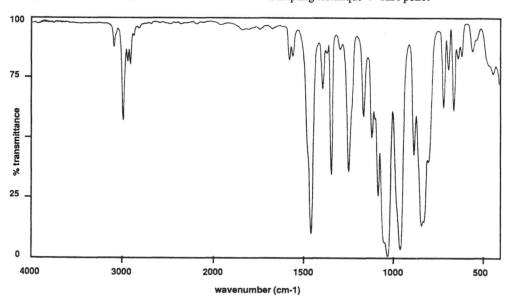

Chemical name : O-(4-bromo-2,5-dichlorophenyl) O,O-diethyl phosphorothioate

Other names : Filariol, Nexagan
Type : acaricide, insecticide
Brutoformula : C10H12BrCl2O3PS   CAS nr : 4824-78-6
Molecular mass : 394.0491   Exact mass : 391.8805704
Instrument : Bruker IFS-85   Optical resolution : 2 cm-1
Scans : 32   Sampling technique : KBr pellet

Band maxima with relative intensity :

553	13	657	38	688	20	715	37	838	87
881	56	956	97	1024	100	1080	74	1116	49
1163	40	1245	63	1342	65	1390	28	1456	90
1575	16	2904	18	2985	42	3089	11		

COMPOUND : **Bromopropylate**

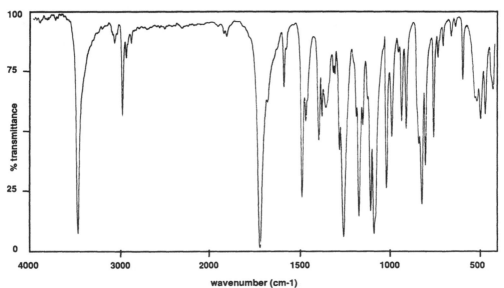

Chemical name   : isopropyl 4,4'-dibromobenzilate

Other names	: Acarol, Neoron 500, GS 19851			
Type	: acaricide			
Brutoformula	: C17H16Br2O3	CAS nr	:	18181-80-1
Molecular mass	: 428.12327	Exact mass	:	425.9467159
Instrument	: Bruker IFS-85	Optical resolution	:	2 cm-1
Scans	: 32	Sampling technique	:	KBr pellet

Band maxima with relative intensity :

423	33	465	43	491	45	588	28	650	10
695	14	724	19	747	53	793	65	814	81
902	49	928	46	982	53	1013	75	1082	94
1101	84	1144	47	1166	86	1254	95	1277	58
1301	26	1353	40	1375	44	1391	54	1464	46
1487	78	1588	31	1718	100	1899	10	2932	18
2978	43	3065	12	3469	93				

41

COMPOUND : **Bromoxynil**

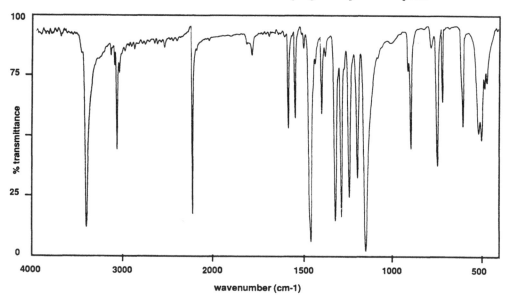

Chemical name	: 3,5-dibromo-4-hydroxybenzonitrile			

Other names : Brominil, Buctril, LH 6950, MB 10731
Type : herbicide
Brutoformula : C7H3Br2NO   CAS nr   : 1689-84-5
Molecular mass : 276.91606   Exact mass   : 274.8582411
Instrument : Bruker IFS-85   Optical resolution : 2 cm-1
Scans : 32   Sampling technique : KBr pellet

Band maxima with relative intensity :

502	53	604	47	718	37	748	64	893	56
1154	100	1200	68	1246	77	1292	85	1326	87
1401	41	1464	95	1500	13	1548	43	1587	47
1784	16	2230	83	3071	56	3096	20	3412	88

42

COMPOUND : **Bronopol**

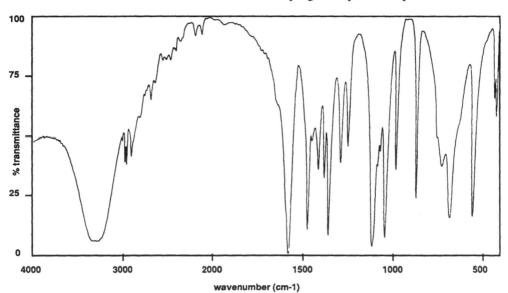

Chemical name : 2-bromo-2-nitro-1,3-propanediol

Other names	: Bronodiol, Bronocot		
Type	: bactericide		
Brutoformula	: C3H6BrNO4	CAS nr	: 52-51-7
Molecular mass	: 199.98957	Exact mass	: 198.9480672
Instrument	: Bruker IFS-85	Optical resolution	: 2 cm-1
Scans	: 32	Sampling technique	: KBr pellet

Band maxima with relative intensity :

412	41	423	33	537	83	661	84	708	62
851	76	966	64	1023	92	1095	96	1235	54
1273	61	1339	92	1365	67	1398	64	1454	89
1558	100	2672	35	2884	59	2936	62	2954	61
3257	95								

COMPOUND : **Buminafos**

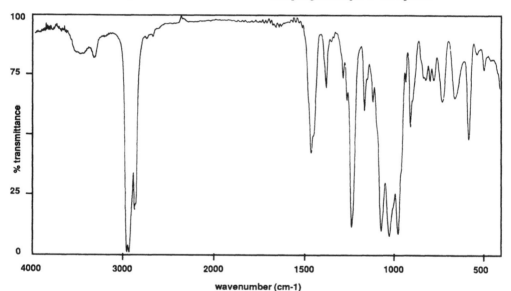

Chemical name : dibutyl [1-(butylamino)cyclohexyl]phosphonate

Other name   : Trakephon
Type   : herbicide
Brutoformula   : C18H38NO3P   CAS nr   : 51249-05-9
Molecular mass : 347.48222   Exact mass   : 347.2589143
Instrument   : Bruker IFS-85   Optical resolution : 2 cm-1
Scans   : 32   Sampling technique : KBr pellet

Band maxima with relative intensity :

402	36	498	22	581	52	657	34	729	36
798	27	821	27	906	46	974	92	1023	93
1071	91	1118	36	1165	40	1238	89	1286	26
1379	30	1462	58	2871	82	2932	100		

COMPOUND : **Bupirimate**

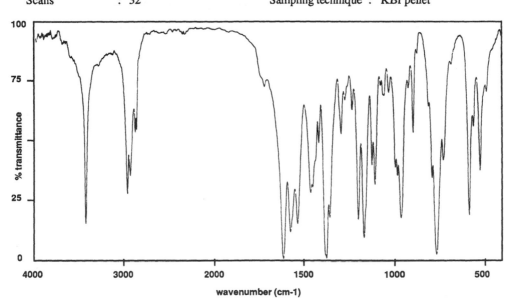

Chemical name   : 5-butyl-2-ethylamino-6-methylpyrimidin-4-yl dimethylsulphamate

Other names   : Nimrod, PP 588, R 70588
Type   : fungicide
Brutoformula   : C13H24N4O3S   CAS nr   : 41483-43-6
Molecular mass   : 316.42123   Exact mass   : 316.1568982
Instrument   : Bruker IFS-85   Optical resolution : 2 cm-1
Scans   : 32   Sampling technique : KBr pellet

Band maxima with relative intensity :

524	62	585	81	728	58	765	98	790	65
899	46	964	82	1029	29	1055	30	1105	68
1122	60	1164	91	1195	83	1230	36	1292	47
1372	99	1416	50	1460	72	1534	85	1574	88
1615	100	2872	45	2926	64	2958	72	3417	84

45

COMPOUND : **Buprofezin**

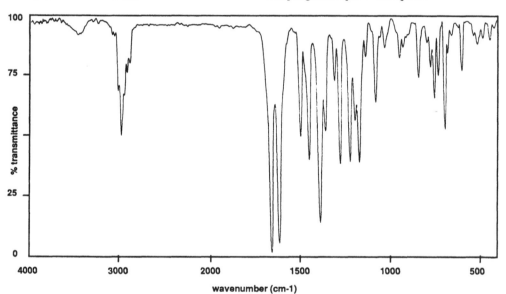

Chemical name   : 2-tert.butylimino-3-isopropyl-5-phenylperhydro-1,3,5-thiadiazinan-4-one

Other names   : Applaud, NNI 750, NN 29285, PP 618
Type   : acaricide, insecticide
Brutoformula   : C16H23N3OS   CAS nr   : 69327-76-0
Molecular mass   : 305.44121   Exact mass   : 305.1561722
Instrument   : Bruker IFS-85   Optical resolution   : 2 cm-1
Scans   : 32   Sampling technique : KBr pellet

Band maxima with relative intensity :

511	12	597	23	694	48	730	25	754	35
775	21	840	26	946	18	1028	13	1082	36
1136	17	1174	62	1195	44	1224	61	1278	62
1307	27	1361	48	1394	87	1452	60	1496	50
1622	96	1664	100	2972	49				

COMPOUND : **Butocarboxim**

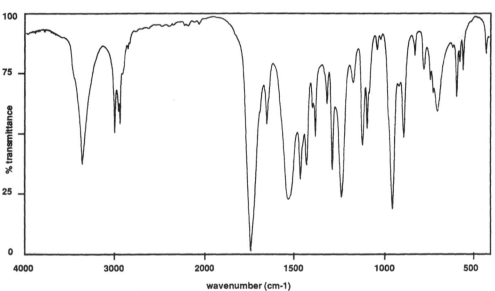

Chemical name   : 3-(methylthio)butanone O-methylcarbamoyloxime

Other names   : Afiline, CO 755, Drawacid A
Type   : insecticide
Brutoformula   : C7H14N2O2S   CAS nr   : 34681-10-2
Molecular mass   : 190.26183   Exact mass   : 190.0775917
Instrument   : Bruker IFS-85   Optical resolution   : 2 cm-1
Scans   : 32   Sampling technique : KBr pellet

Band maxima with relative intensity :

418	16	545	23	565	19	582	34	690	40
730	27	764	23	818	17	881	52	945	82
1028	14	1083	48	1109	55	1160	28	1225	77
1274	65	1303	37	1368	51	1416	63	1452	69
1518	78	1641	46	1731	100	2917	46	2934	40
2976	50	3331	63						

COMPOUND : **Butonate**

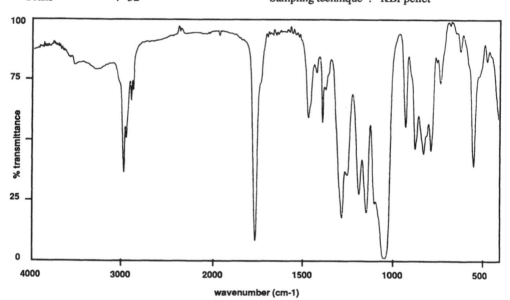

Chemical name   : dimethyl-1-butyryloxy-2,2,2-trichloroethylphosphonate

Other names   : -
Type   : insecticide
Brutoformula   : C8H14Cl3O5P   CAS nr   :
Molecular mass   : 327.53054   Exact mass   : 325.9644382
Instrument   : Bruker IFS-85   Optical resolution : 2 cm-1
Scans   : 32   Sampling technique : KBr pellet

Band maxima with relative intensity :

547	61	617	12	730	26	785	54	825	55
871	53	923	44	1039	100	1143	80	1184	72
1278	82	1383	42	1461	40	1763	92	2878	32
2963	63								

COMPOUND : **Butoxycarboxim**

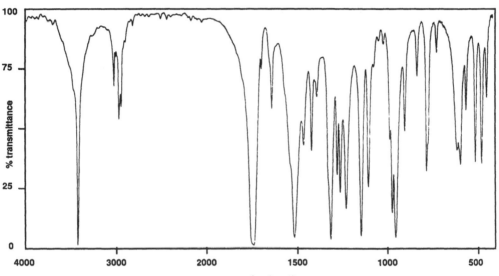

Chemical name   : 3-(methylsulphonyl)butanone O-methylcarbamoyloxime

Other names   : Plant Pin
Type   : insecticide
Brutoformula   : C7H14N2O4S
Molecular mass   : 222.26063
Instrument   : Bruker IFS-85
Scans   : 32

CAS nr   : 34681-23-7
Exact mass   : 222.0674199
Optical resolution : 2 cm-1
Sampling technique : KBr pellet

Band maxima with relative intensity :

439	37	467	66	503	65	555	43	584	66
716	18	771	69	824	28	892	52	940	97
959	86	1095	75	1135	96	1217	84	1250	78
1268	70	1301	97	1381	37	1410	60	1455	57
1504	96	1635	42	1733	100	2928	41	2953	46
3008	32	3403	99						

COMPOUND : **Butylate**

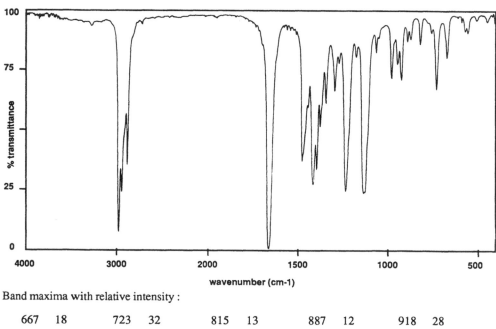

Chemical name   : S-ethyl diisobutylthiocarbamate

Other name   : Sutan, R 1910
Type   : herbicide
Brutoformula   : C11H23NOS   CAS nr   : 2008-41-5
Molecular mass   : 217.37206   Exact mass   : 217.1500258
Instrument   : Bruker IFS-85   Optical resolution   : 2 cm-1
Scans   : 32   Sampling technique   : KBr pellet

Band maxima with relative intensity :

667	18	723	32	815	13	887	12	918	28
941	21	972	27	10564	16	1124	76	1224	75
1290	33	1336	38	1369	48	1386	66	1406	72
1467	62	1650	100	1959	10	2871	64	2929	76
2960	93								

COMPOUND : **Captafol**

Chemical name : 1,2,3,6-tetrahydro-N-(1,1,2,2-tetrachloroethylthio)phthalimide

Other name	: Haipen, Merpafol, Difolatan		
Type	: fungicide		
Brutoformula	: C10H9ClNO2S	CAS nr	: 2425-06-1
Molecular mass	: 349.06073	Exact mass	: 346.9108079
Instrument	: Bruker IFS-85	Optical resolution	: 2 cm-1
Scans	: 32	Sampling technique	: KBr pellet

Band maxima with relative intensity :

420	24	567	42	627	22	688	58	712	51
763	63	810	39	879	39	899	28	1001	26
1046	29	1063	43	1138	75	1195	51	1263	51
1282	82	1313	40	1346	31	1443	28	1725	100
2952	33	2983	38						

COMPOUND : **Captan**

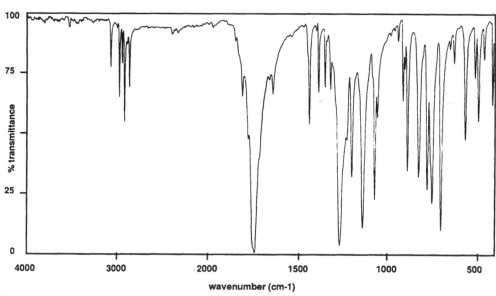

Chemical name   : 1,2,3,6-tetrahydro-N-(trichloromethylthio)phthalimide

Other names   : Pillarcap, Merpan, Orthocide
Type   : fungicide
Brutoformula   : C9H8Cl3NO2D
Molecular mass   : 300.58861
Instrument   : Bruker IFS-85
Scans   : 32

CAS nr   : 133-06-2
Exact mass   : 298.9341302
Optical resolution : 2 cm-1
Sampling technique : KBr pellet

Band maxima with relative intensity :

453	18	485	44	503	25	556	52	617	19
690	90	739	78	766	73	813	67	876	65
901	35	928	10	1040	42	1056	77	1126	89
1187	67	1256	96	1307	30	1339	29	1377	31
1430	45	1634	32	1734	100	1802	33	2844	30
2897	44	2924	19	2956	34	3054	21		

COMPOUND : **Carbaryl**

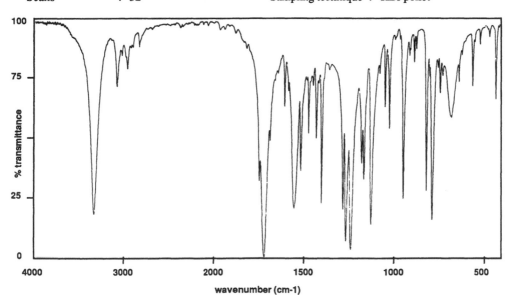

Chemical name   : 1-naphthyl methylcarbamate

Other name   : Murvin, Patrin, Dicarbam, Sevin
Type   : insecticide, growth regulator
Brutoformula   : C12H11NO2   CAS nr   : 63-25-2
Molecular mass   : 210.22697   Exact mass   : 201.078972
Instrument   : Bruker IFS-85   Optical resolution : 2 cm-1
Scans   : 32   Sampling technique : KBr pellet

Band maxima with relative intensity :

422	32	510	9	552	27	627	25	667	40
729	30	774	84	806	71	863	11	875	17
901	14	933	75	1011	45	1036	36	1113	86
1153	66	1167	60	1224	96	1251	93	1269	79
1387	77	1417	49	1461	47	1505	63	1542	79
1597	36	1710	100	1739	67	2810	10	2940	20
3059	27	3310	82						

COMPOUND : **Carbendazim**

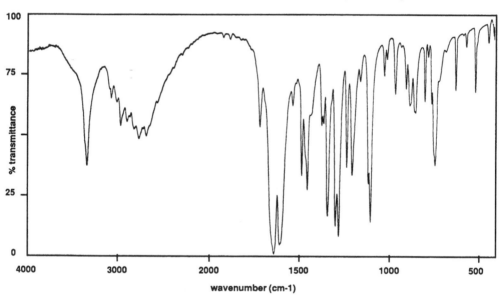

Chemical name   : methyl benzimidazol-2-ylcarbamate

Other names   : Derosal, BCM, Carbendazol, Mergal
Type   : fungicide
Brutoformula   : C9H9N3O2   CAS nr   : 10605-21-7
Molecular mass   : 191.19098   Exact mass   : 191.0694692
Instrument   : Bruker IFS-85   Optical resolution : 2 cm-1
Scans   : 32   Sampling technique : KBr pellet

Band maxima with relative intensity :

435	10	508	30	559	11	619	30	733	62
792	34	844	39	876	36	896	29	956	32
1016	24	1095	86	1195	66	1225	63	1268	92
1286	88	1329	83	1363	45	1443	72	1475	66
1525	37	1598	96	1629	100	1711	46	2751	51
2949	45	3057	34	3325	62				

COMPOUND : **Carbetamide**

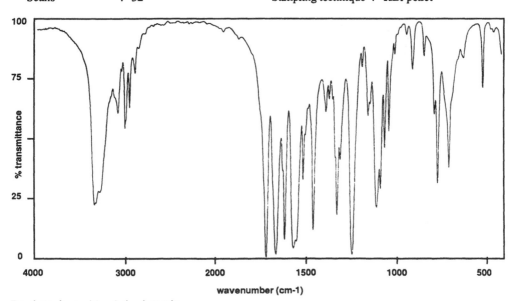

Chemical name : (R)-(-)-1-ethylcarbamoyl)ethyl phenylcarbamate

Other names	: Legurame, Carbetamex			
Type	: herbicide			
Brutoformula	: C12H16N2O3		CAS nr	: 16118-49-3
Molecular mass	: 236.27292		Exact mass	: 236.1160823
Instrument	: Bruker IFS-85		Optical resolution	: 2 cm-1
Scans	: 32		Sampling technique	: KBr pellet

Band maxima with relative intensity :

508	28	693	62	757	68	775	39	833	14
899	20	1029	46	1052	53	1075	70	1096	78
1146	40	1179	19	1232	98	1315	82	1359	32
1377	38	1445	88	1501	67	1554	96	1603	92
1650	98	1704	100	2875	22	2936	36	2985	45
3064	39	3316	78						

COMPOUND : **Carbofuran**

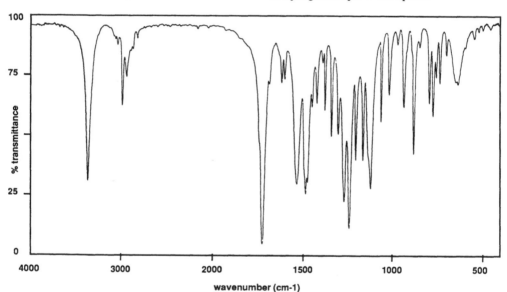

Chemical name   : 2,3-dihydro-2,2-dimethylbenzofuran-7-yl methylcarbamate

Other names   : Furadan, Curaterr, Bayer 70143, NIA 10242
Type   : acaricide, insecticide, nematicide
Brutoformula   : C12H15NO3   CAS nr   : 1563-66-2
Molecular mass   : 221.25825   Exact mass   : 221.1051845
Instrument   : Bruker IFS-85   Optical resolution   : 2 cm-1
Scans   : 32   Sampling technique : KBr pellet

Band maxima with relative intensity :

631	31	697	19	733	31	754	28	771	45
792	39	875	61	929	41	967	15	1010	36
1056	47	1113	76	1157	64	1197	63	1231	93
1260	81	1296	53	1334	54	1372	43	1417	40
1444	41	1477	78	1528	74	1600	30	1618	31
1719	100	2932	29	2978	41	3362	73		

COMPOUND : **Carbophenothion**

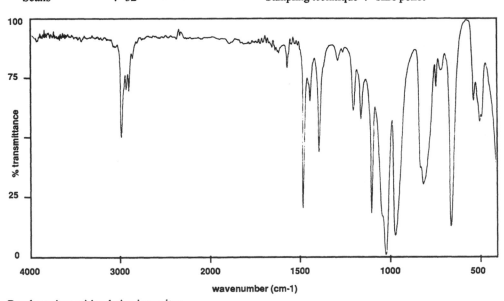

Chemical name	: S-(4-chlorophenylthio)methyl O,O-diethyl phosphorodithioate

Other name   : Trithion, R 1303, Garrathion

Type	: acaricide, insecticide		
Brutoformula	: C11H16ClO2PS3	CAS nr	: 786-19-6
Molecular mass	: 342.85573	Exact mass	: 341.973857
Instrument	: Bruker IFS-85	Optical resolution	: 2 cm-1
Scans	: 32	Sampling technique	: KBr pellet

Band maxima with relative intensity :

501	43	535	34	654	87	746	28	810	70
961	91	1011	100	1095	82	1160	42	1203	38
1291	17	1388	56	1441	134	1476	80	1573	20
2900	30	2934	30	2981	50				

COMPOUND : **Carbosulfan**

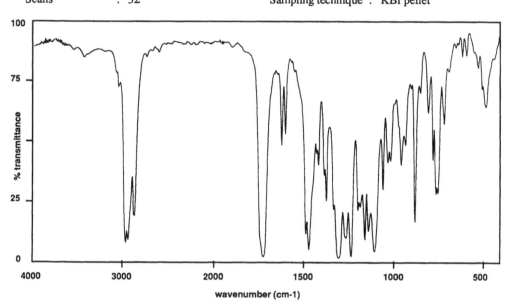

Chemical name	:	2,3-dihydro-2,2-dimethyl-7-benzofuranyl[(dibutylamino)thio]-methylcarbamate
Other names	:	Marshal, Posse, FMC 35001, MEX 212
Type	:	insecticide
Brutoformula	:	C20H32N2O3S
Molecular mass	:	380.54964
Instrument	:	Bruker IFS-85
Scans	:	32

CAS nr	: 55285-14-8
Exact mass	: 380.2133486
Optical resolution	: 2 cm-1
Sampling technique	: KBr pellet

Band maxima with relative intensity :

481	36	589	14	613	14	713	43	759	73
777	58	803	38	877	84	954	60	1027	59
1055	71	1104	97	1136	88	1157	92	1194	79
1233	99	1260	91	1301	100	1370	75	1412	60
1466	96	1483	89	1598	47	1620	52	1723	99
1958	16	2862	82	2957	93				

58

COMPOUND : **Carboxin**

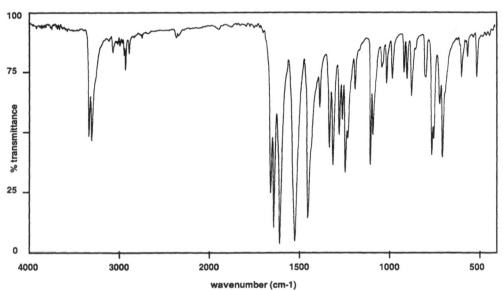

Chemical name   : 5,6-dihydro-2-methyl-1,4-oxathiin-3-carboxanilide

Other names   : DCMO, Vitavax
Type   : fungicide
Brutoformula   : C12H13NO2S   CAS nr   : 5234-68-4
Molecular mass   : 235.30291   Exact mass   : 235.0666939
Instrument   : Bruker IFS-85   Optical resolution : 2 cm-1
Scans   : 32   Sampling technique : KBr pellet

Band maxima with relative intensity :

506	30	557	17	591	23	692	60	711	35
751	60	787	31	868	37	894	30	910	27
974	30	1004	32	1032	25	1079	54	1092	66
1180	34	1231	69	1250	47	1267	53	1301	66
1322	59	1376	42	1437	89	1508	98	1594	100
1629	93	1648	78	2876	20	2917	27	3056	19
3287	57	3319	55						

59

COMPOUND : **Chinomethionate**

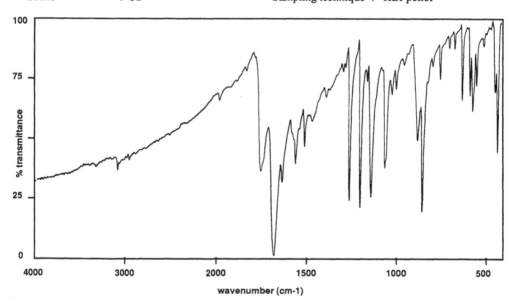

Chemical name   : 6-methyl-1,3-dithiolo[4,5-b] quinoxalin-2-one

Other name   : Chinomethionate, Oxythiochinox, Morestan, SS 2074
Type   : acaricide, fungicide
Brutoformula   : C10H6N2OS2   CAS nr   : 2439-01-2
Molecular mass   : 234.29212   Exact mass   : 233.9921535
Instrument   : Bruker IFS-85   Optical resolution : 2 cm-1
Scans   : 32   Sampling technique : KBr pellet

Band maxima with relative intensity :

422	55	436	30	538	27	560	38	575	31
618	33	661	12	690	11	739	24	838	81
865	50	984	29	1007	31	1047	62	1122	74
1183	79	1243	76	1492	53	1542	61	1617	68
1661	100	1736	64	3050	64				

COMPOUND : **3-(3-Chloroanilo)-perhydrofuran-2-on**

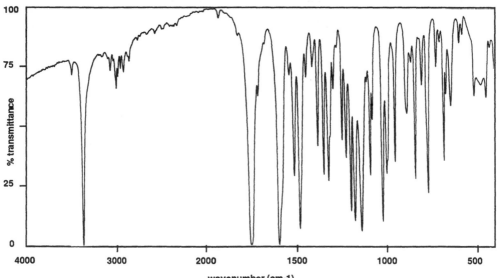

Chemical name   : 3-(3-chloroanilo)-perhydrofuran-2-on

Other name   :
Type   : metabolite of Cyprofuram
Brutoformula   : C10H10ClNO2   CAS nr   : -
Molecular mass   : 211.6497   Exact mass   : 211.040000
Instrument   : Bruker IFS-85   Optical resolution : 2 cm-1
Scans   : 32   Sampling technique : KBr pellet

Band maxima with relative intensity :

447	39	513	38	595	15	641	42	668	37
679	65	721	25	768	79	802	33	839	73
886	45	952	65	999	70	1022	90	1078	47
1090	71	1141	94	1179	90	1200	86	1226	63
1248	56	1298	34	1325	73	1352	70	1385	58
1414	25	1449	29	1486	93	1515	71	1541	28
1599	100	1756	99	2921	26	2955	24	3005	33
3071	25	3375	99						

COMPOUND : **Chlorbromuron**

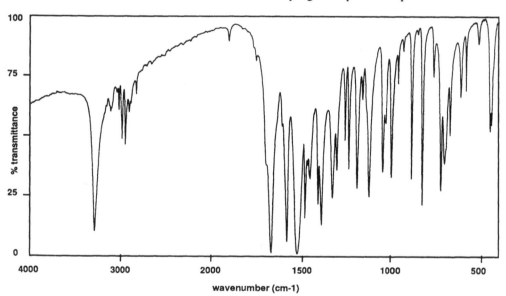

Chemical name   : 3-(4-bromo-3-chlorophenyl)-1-methoxy-1-methylurea

Other names   : Maloran, C 6313
Type   : herbicide
Brutoformula   : C9H10BrClN2O2   CAS nr   : 13360-45-7
Molecular mass   : 293.54925   Exact mass   : 291.9614637
Instrument   : Bruker IFS-85   Optical resolution : 2 cm-1
Scans   : 32   Sampling technique : KBr pellet

Band maxima with relative intensity :

443	48	508	10	575	30	604	33	662	49
694	61	712	73	754	24	816	79	875	68
920	13	949	27	983	67	1032	65	1111	75
1149	34	1179	72	1225	63	1247	51	1294	64
1319	76	1382	87	1401	78	1445	67	1474	84
1518	100	1574	94	1662	99	1896	9	2811	32
2892	39	2933	53	2970	51	3007	39	3100	39
3275	90								

COMPOUND : **Chlorbufam**

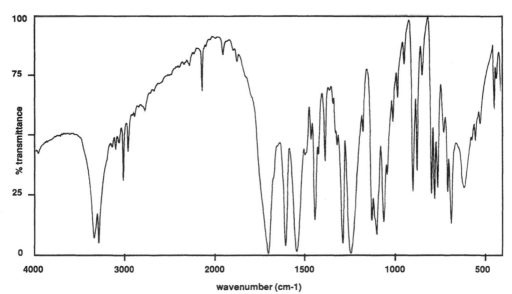

Chemical name : 1-methylprop-2-ynyl 3-chlorophenylcarbamate

Other names	: BIPC		
Type	: herbicide		
Brutoformula	: C11H10ClNO2	CAS nr	: 1967-16-4
Molecular mass	: 223.66085	Exact mass	: 233.0400005
Instrument	: Bruker IFS-85	Optical resolution	: 2 cm-1
Scans	: 32	Sampling technique	: KBr pellet

Band maxima with relative intensity :

438	38	541	52	605	72	677	87	698	74
719	48	752	72	769	77	786	74	838	24
866	65	890	73	937	19	973	33	999	44
1050	86	1089	92	1114	86	1163	48	1231	100
1274	95	1374	60	1432	85	1452	51	1533	99
1596	96	1694	99	1943	15	2119	30	2939	56
2991	68	3074	55	3260	94	3309	92		

COMPOUND : **Chlordane (cis)**

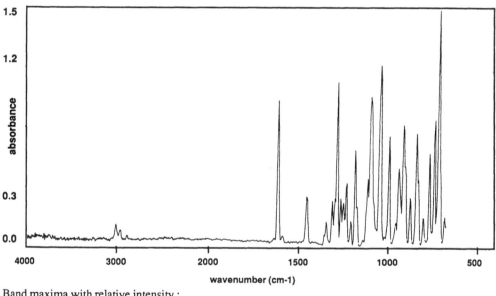

Chemical name	:	1,2,4,5,6,7,8,8-octachloro-2,3,3a,4,7,7a-hexahydro-4,7-methanoindene	

Other names   : Belt, Chlortox, Intox, Octachlor
Type   : insecticide
Brutoformula   : C10H6Cl8   CAS nr   : 5103-71-9
Molecular mass : 409.78332   Exact mass   : 405.7977724
Instrument   : Biorad Tracer GC/FTIR   Optical resolution : 2 cm-1
Scans   : 256   Sampling technique : cryotrapping

Band maxima with relative intensity :

692	35	703	100	727	52	755	43	818	25
828	44	863	17	889	27	902	45	928	27
979	42	1024	70	1030	45	1074	53	1082	60
1157	16	1168	42	1217	29	1270	69	1297	18
1441	17	1604	52						

COMPOUND : **Chlordane(trans)**

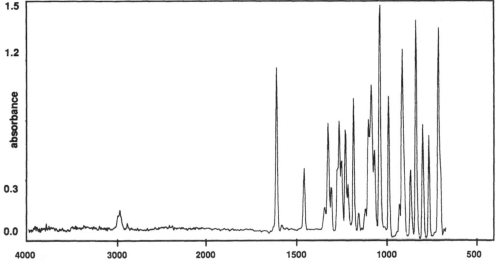

Chemical name   : 1,2,4,5,6,7,8,8-octachloro-2,3,3a,4,7,7a-hexahydro-4,7-methanoindene

Other names   : Belt, Chlortox, Intox, Octachlor
Type   : insecticide
Brutoformula   : C10H6Cl8   CAS nr   : 5103-74-2
Molecular mass : 409.78332   Exact mass   : 405.7977724
Instrument   : Biorad Tracer GC/FTIR   Optical resolution   : 2 cm-1
Scans   : 256   Sampling technique : cryotrapping

Band maxima with relative intensity :

689	47	699	89	752	46	786	43	824	89
858	28	893	36	903	72	979	65	1027	100
1058	33	1075	57	1092	45	1176	61	1208	18
1221	49	1244	28	1256	49	1267	26	1302	20
1319	41	1451	26	1604	65				

COMPOUND : **Chlordimeform (HCl)**

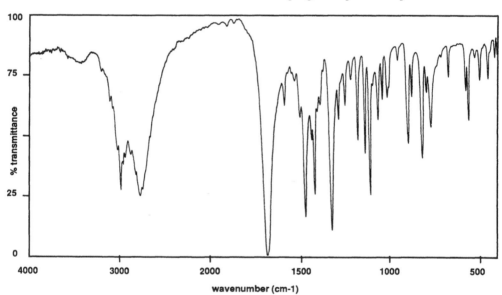

Chemical name : N'-(4-chloro-2-methylphenyl)-N,N-dimethylmethanimidamide (HCl)

Other names : Schering 36268, Sch 1066, Fundal, Chlorphenamidine
Type : acaricide / insecticide
Brutoformula : C10H14Cl2N2   CAS nr : 19780-95-9
Molecular mass : 233.1   Exact mass : 233.14248
Instrument : Bruker IFS-85   Optical resolution : 2 cm-1
Scans : 32   Sampling technique : KBr pellet

Band maxima with relative intensity :

416	18	456	27	501	27	565	44	578	31
674	26	776	47	799	32	826	60	880	34
902	53	955	18	1014	34	1043	35	1069	43
1117	75	1143	57	1186	52	1221	26	1255	37
1293	43	1338	89	1433	74	1488	83	1597	36
1704	100	2792	73	3007	71				

COMPOUND : **Chlordimeform**

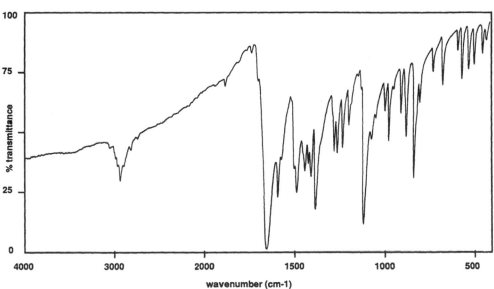

Chemical name	: N'-(4-chloro-2-methylphenyl)-N,N-dimethylmethanimidamide		
Other names	: Schering 36268, Sch 1066, Fundal, Chlorphenamidine		
Type	: acaricide / insecticide		
Brutoformula	: C10H13ClN2	CAS nr	: 6164-98-3
Molecular mass	: 196.68151	Exact mass	: 196.0767193
Instrument	: Bruker IFS-85	Optical resolution	: 2 cm-1
Scans	: 32	Sampling technique	: KBr pellet

Band maxima with relative intensity :

454	15	497	20	529	22	565	26	589	14
672	29	726	23	796	36	826	68	871	51
902	41	968	53	991	40	1101	88	1190	46
1223	56	1253	58	1270	58	1368	82	1395	68
1430	66	1474	75	1580	77	1637	100	2915	71

COMPOUND : **Chlorfenprop-methyl**

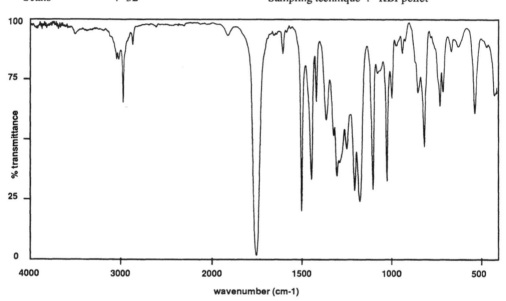

Chemical name   : methyl 2-chloro-3-(4-chlorophenyl)propionate

Other names	: Methachlorphenprop, Bidisin		
Type	: herbicide		
Brutoformula	: C10H10Cl2O2	CAS nr	: 14437-17-3
Molecular mass	: 233.096	Exact mass	: 232.0057804
Instrument	: Bruker IFS-85	Optical resolution	: 2 cm-1
Scans	: 32	Sampling technique	: KBr pellet

Band maxima with relative intensity :

418	32	527	39	657	13	703	30	720	36
810	54	843	31	929	14	986	33	1016	68
1095	72	1167	77	1196	72	1240	54	1296	66
1353	42	1409	34	1437	67	1492	81	1597	14
1748	100	2846	10	2954	34				

68

COMPOUND : **Chlorfenvinfos   (cis)**

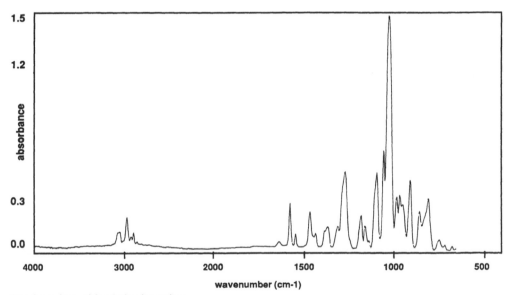

Chemical name   : 2-chloro-1-(2,4-dichlorophenyl)vinyl diethyl phosphate (cis)

Other name   : Birlane, Sapecron, Apachlor
Type   : insecticide, acaricide
Brutoformula   : C12H14Cl3O4P   CAS nr   : 470-90-6
Molecular mass : 359.57574   Exact mass   : 357.9695241
Instrument   : Biorad Tracer GC/FTIR Optical resolution : 2 cm-1
Scans   : 256   Sampling technique : cryotrapping

Band maxima with relative intensity :

812 24	862 18	913 32	955 20	1032 100
1059 47	1100 28	1110 28	1189 17	1278 34
1316 10	1474 17	1586 23	2985 13	

COMPOUND : **Chlofenvinphos (trans)**

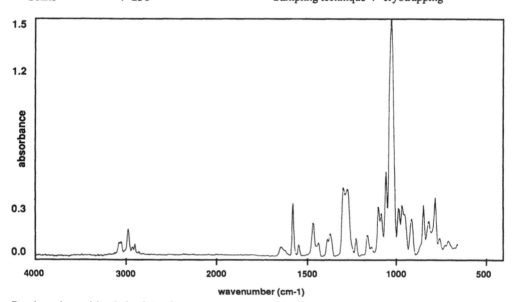

Chemical name   :   2-chloro-1-(2,4-dichlorophenyl)vinyl diethyl phosphate (trans)

Other names   :   Birlane, Sapecron, Apachlor, Birlane
Type   :   insecticide / acaricide
Brutoformula   :   C12H14Cl3O4P   CAS nr   :   470-90-6
Molecular mass   :   359.57574   Exact mass   :   357.9695241
Instrument   :   Biorad Tracer GC/FTIR Optical resolution :   2 cm-1
Scans   :   256   Sampling technique :   cryotrapping

Band maxima with relative intensity :

787	22	852	20	919	15	959	17	1024	100
1060	36	1105	22	1279	26	1303	28	1471	14
1586	25	2985	11						

COMPOUND : **Choridazon**

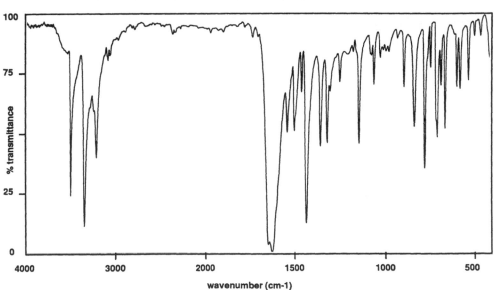

Chemical name : 5-amino-4-chloro-2-phenylpyridazin-3-one

Other names : Pyrazon, Pyramin
Type : herbicide
Brutoformula : C10H8ClN3O   CAS nr : 1698-60-8
Molecular mass : 221.64776   Exact mass : 221.0355836
Instrument : Bruker IFS-85   Optical resolution : 2 cm-1
Scans : 32   Sampling technique : KBr pellet

Band maxima with relative intensity :

456	8	491	8	524	27	572	31	590	30
656	47	679	29	699	51	737	22	768	64
828	46	886	30	1019	17	1054	29	1137	54
1241	28	1309	53	1347	55	1422	87	1451	32
1491	48	1531	49	1614	100	3060	18	3189	59
3320	88	3471	75						

COMPOUND : **Chlormephos**

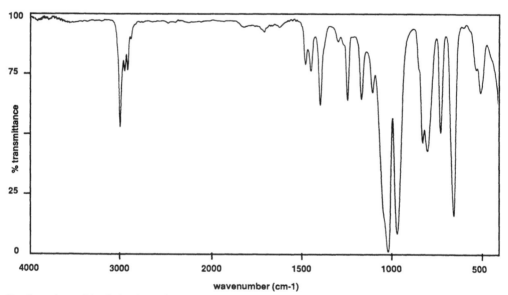

Chemical name   : S-(chloromethyl) O,O-diethylphosphorodithionate

Other names   : Dotan
Type   : insecticide
Brutoformula   : C5H12ClO2PS2   CAS nr   : 24934-91-6
Molecular mass   : 234.69695   Exact mass   : 233.9704859
Instrument   : Bruker IFS-85   Optical resolution : 2 cm-1
Scans   : 32   Sampling technique : KBr pellet

Band maxima with relative intensity :

505	32	653	84	725	49	798	57	825	53
964	92	1012	100	1099	32	1159	35	1236	35
1390	37	1442	23	1473	20	2902	22	2985	46

COMPOUND : **Chlormequat chloride**

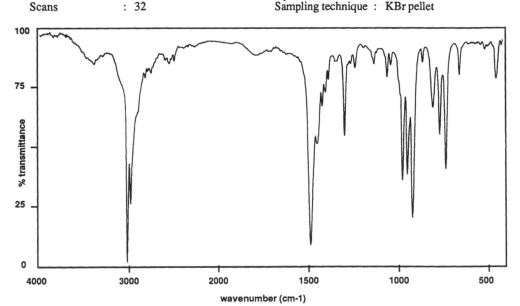

Chemical name : 2-chloroethyltrimethylammonium chloride

Other names : Cycocel, CCC, Lihocin
Type : growth regulator
Brutoformula : C5H13Cl2N   CAS nr : 991-81-5
Molecular mass : 158.07206   Exact mass : 157.0424992
Instrument : Bruker IFS-85   Optical resolution : 2 cm-1
Scans : 32   Sampling technique : KBr pellet

Band maxima with relative intensity :

458	22	659	21	732	61	767	46	806	34
865	15	919	81	949	63	976	65	1064	22
1139	16	1245	18	1301	46	1423	34	1485	93
2974	75	3009	100						

COMPOUND : **Chlorobenzilate**

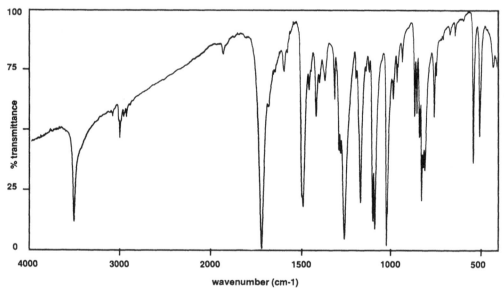

Chemical name   : ethyl 4,4'-dichlorobenzilate

Other names   : Akar, Benzilan, Gesaspint
Type   : acaricide
Brutoformula   : C16H14Cl2O3   CAS nr   : 510-15-6
Molecular mass   : 325.19418   Exact mass   : 324.0319929
Instrument   : Bruker IFS-85   Optical resolution : 2 cm-1
Scans   : 32   Sampling technique : KBr pellet

Band maxima with relative intensity :

500	53	534	64	634	10	750	44	801	67
820	79	828	52	834	53	847	43	859	44
926	21	956	30	975	37	1014	98	1082	91
1092	88	1163	80	1253	96	1273	59	1282	58
1307	37	1362	29	1411	44	1450	33	1484	82
1590	25	1715	100	2983	53	3486	88		

74

COMPOUND : **4-Chlorophenoxy-acetic acid**

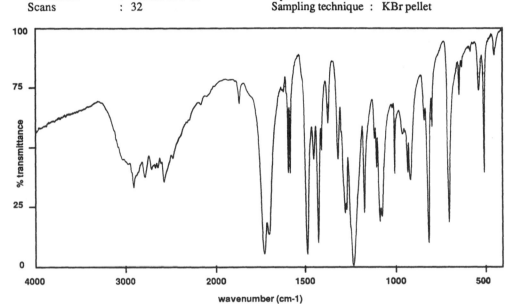

Chemical name	: 4-chlorophenoxy-acetic acid		
Other name	: 4-CPA		
Type	: growth regulator		
Brutoformula	: C8H7O3Cl	CAS nr	: 122-88-3
Molecular mass	: 186.591619	Exact mass	: 186.0083676
Instrument	: Bruker IFS-85	Optical resolution	: 2 cm-1
Scans	: 32	Sampling technique	: KBr pellet

Band maxima with relative intensity :

448	11	506	61	535	26	646	27	707	82
797	41	820	90	840	38	920	63	934	60
1004	61	1088	81	1105	58	1173	77	1236	100
1278	77	1316	54	1371	39	1408	50	1429	89
1452	54	1489	94	1584	60	1595	60	1735	94
1872	30	2580	63	2787	61	2911	65		

COMPOUND : **Chlorthalonil**

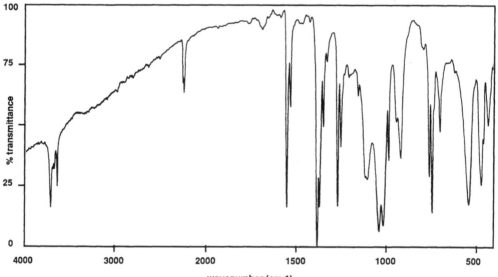

Chemical name	: tetrachloroisophtalonitrile		
Other names	: Bravo, Daconil, TPN		
Type	: fungicide		
Brutoformula	: C8Cl4N2	CAS nr	: 1897-45-6
Molecular mass	: 265.9146	Exact mass	: 263.8815588
Instrument	: Bruker IFS-85	Optical resolution	: 2 cm-1
Scans	: 32	Sampling technique	: KBr pellet

Band maxima with relative intensity :

432	48	471	74	540	82	695	51	738	86
755	70	914	62	981	63	1009	91	1032	93
1102	71	1247	58	1266	83	1345	49	1366	83
1379	100	1527	41	1549	83	2240	34	3619	74
3694	83								

COMPOUND : **Chloroxuron**

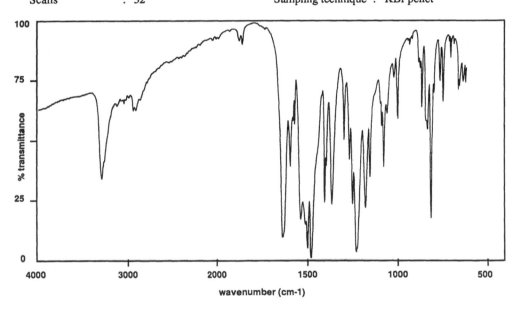

Chemical name   : 3-[4-(4-chlorophenoxy)phenyl]-1,1-dimethylurea

Other names   : Chloroxifenidim, Chlorphencarb, Tenoran, Norex
Type   : herbicide
Brutoformula   : C15H15ClN2O2   CAS nr   : 1982-47-4
Molecular mass : 290.752   Exact mass   : 290.082196
Instrument   : Bruker IFS-85   Optical resolution   : 2 cm-1
Scans   : 32   Sampling technique : KBr pellet

Band maxima with relative intensity :

643	24	670	28	712	14	755	33	771	24
821	83	841	45	874	35	1007	40	1029	23
1086	61	1098	43	1163	65	1187	78	1235	97
1258	77	1276	58	1305	49	1373	77	1413	76
1485	100	1506	95	1546	83	1582	43	1606	60
1647	91	1873	9	2957	37	3302	66		

COMPOUND : **Chlorpropham**

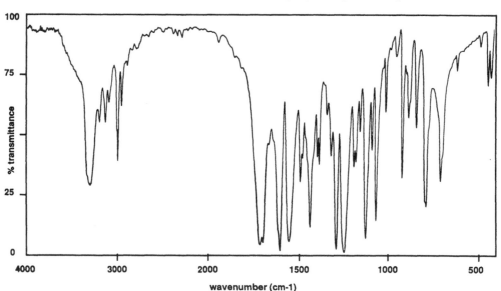

Chemical name	: isopropyl 3-chlorophenylcarbamate		
Other names	: Furloe, Pommetrol, CIPC, Chloro-IPC, Chlor-IFC		
Type	: herbicide, growth regulator		
Brutoformula	: C10H12ClNO2	CAS nr	: 101-21-3
Molecular mass	: 213.66564	Exact mass	: 213.0556497
Instrument	: Bruker IFS-85	Optical resolution	: 2 cm-1
Scans	: 32	Sampling technique	: KBr pellet

Band maxima with relative intensity :

438	29	605	22	693	69	772	80	827	47
869	42	904	68	940	16	997	40	1051	86
1076	56	1110	93	1144	48	1179	63	1232	100
1277	98	1307	59	1375	62	1385	59	1425	89
1480	70	1542	95	1593	99	1704	97	2939	38
2978	61	3123	45	3189	45	3290	72		

COMPOUND : **Chlorpyrifos**

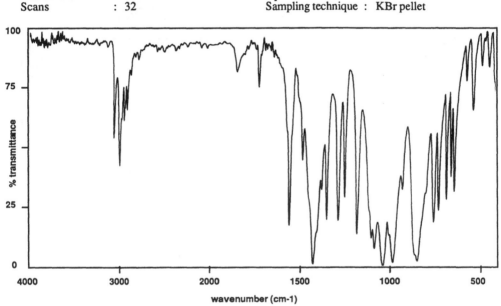

Chemical name   : O,O-diethyl O-(3,5,6-trichloro-2-pyridyl)phosphorothioate

Other names   : Lorsban, Detmol, Loxiran, Dursban
Type   : insecticide
Brutoformula   : C9H11Cl3NO3PS   CAS nr   : 2921-88-2
Molecular mass   : 350.58568   Exact mass   : 348.9262821
Instrument   : Bruker IFS-85   Optical resolution : 2 cm-1
Scans   : 32   Sampling technique : KBr pellet

Band maxima with relative intensity :

441	15	481	14	529	33	567	21	632	68
649	62	675	71	718	76	745	81	834	97
918	67	967	98	1021	100	1069	92	1169	86
1239	71	1274	81	1339	80	1411	99	1474	55
1549	83	1724	24	1848	18	2905	34	2938	39
2984	58	3051	46						

COMPOUND : **Chlorpyrifos-methyl**

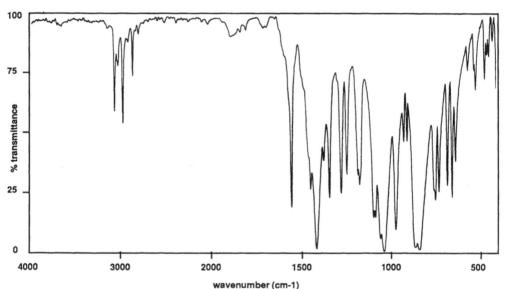

Chemical name   : O,O-dimethyl O-(3,5,6-trichloro-2-pyridyl)phosphorothioate

Other names   : Reldan, Tumar, Trichlormethylfos
Type   : insecticide, acaricide
Brutoformula   : C7H7Cl3NO3PS   CAS nr   : 5598-13-0
Molecular mass   : 322.5315   Exact mass   : 320.8949837
Instrument   : Bruker IFS-85   Optical resolution : 2 cm-1
Scans   : 32   Sampling technique : KBr pellet

Band maxima with relative intensity :

427	11	446	17	468	27	518	31	563	23
628	62	649	77	675	72	721	74	742	78
832	99	901	53	919	53	963	90	1029	100
1091	85	1170	71	1242	67	1273	75	1339	77
1370	61	1410	98	1443	73	1549	81	1881	8
2848	24	2954	44	3049	39				

80

COMPOUND : **Chlorthal-dimethyl**

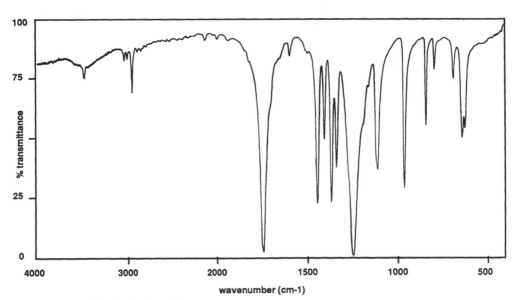

Chemical name : dimethyl tetrachloroterephthalate

Other names : Chlortal-methyl, DCPA, Dacthal
Type : herbicide
Brutoformula : C10H6Cl4O4
Molecular mass : 331.96892
Instrument : Bruker IFS-85
Scans : 32

CAS nr : 1861-32-1
Exact mass : 329.9020164
Optical resolution : 2 cm-1
Sampling technique : KBr pellet

Band maxima with relative intensity :

632	49	681	24	787	20	833	44	956	71
1105	63	1242	100	1335	62	1363	77	1404	50
1441	77	1598	14	1739	98	2954	30	3464	24

COMPOUND : **Chlorthiamid**

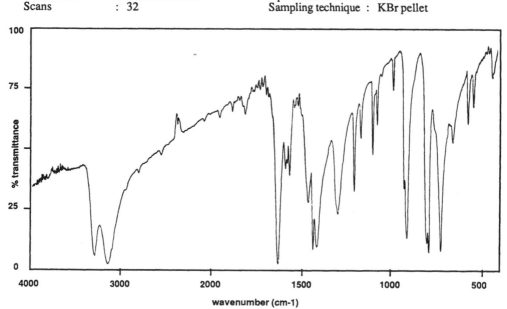

Chemical name   : 2,6-dichlorothiobenzamide

Other name   : Prefix, WL 5792
Type   : herbicide
Brutoformula : C7H5Cl2NS   CAS nr   : 1918-13-4
Molecular mass : 206.0906   Exact mass   : 204.9519751
Instrument   : Bruker IFS-85   Optical resolution : 2 cm-1
Scans   : 32   Sampling technique : KBr pellet

Band maxima with relative intensity :

432	21	534	33	564	40	644	48	712	94
778	95	896	89	972	26	1060	41	1087	53
1154	46	1193	69	1287	78	1407	92	1429	93
1456	73	1560	62	1582	59	1625	100	1804	36
3127	99	3275	96						

COMPOUND : **Chlorthiophos sulfon**

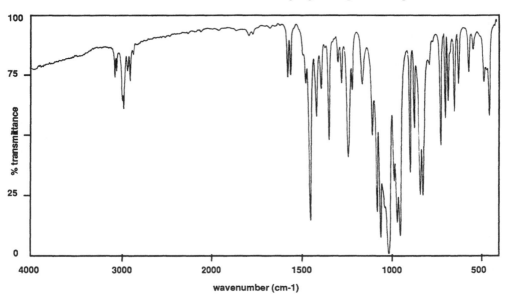

Chemical name   : O,O-diethyl O-[dichloro(methylthio)phenyl] phosphate

Other names   : -
Type   : metabolite of Chlorothiophos
Brutoformula   : C11H15Cl2O5PS2   CAS nr   : -
Molecular mass : 393.2389   Exact mass   : 391.947555
Instrument   : Bruker IFS-85   Optical resolution : 2 cm-1
Scans   : 32   Sampling technique : KBr pellet

Band maxima with relative intensity :

450	41	479	27	538	13	563	23	619	28
642	39	677	35	693	42	720	54	823	75
838	75	870	46	894	65	953	92	970	86
984	69	1019	100	1063	93	1082	82	1107	49
1162	27	1217	30	1242	58	1276	27	1296	18
1348	51	1390	29	1417	41	1455	85	1476	27
1559	23	1575	24	2907	25	2937	21	2980	37
3057	21	3077	23						

COMPOUND : **Chlorthiophos sulfoxide**

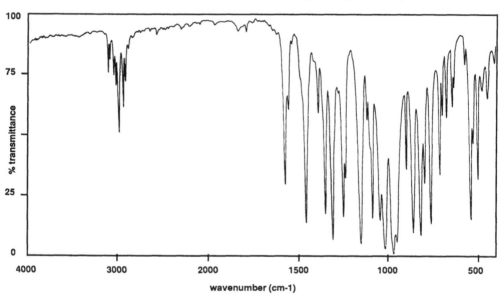

Chemical name   : O,O-diethyl O-[dichloro(methylthio)phenyl] phosphonate

Other name   : -
Type   : metabolite of Chlorothiophos
Brutoformula   : C11H15Cl2O4PS2   CAS nr   : -
Molecular mass   : 377.2395   Exact mass   : 375.95264
Instrument   : Bruker IFS-85   Optical resolution : 2 cm-1
Scans   : 32   Sampling technique : KBr pellet

Band maxima with relative intensity :

447	34	476	31	5015	68	526	53	541	85
568	20	637	38	669	42	692	41	709	66
761	87	795	70	819	92	861	91	895	64
971	100	1017	97	1043	85	1085	84	1114	43
1155	95	1239	67	1252	84	1314	93	1353	82
1390	40	1463	86	1556	38	1578	70	1787	5
2907	25	2932	36	2981	47	3010	27	3034	22
3077	16	3095	21						

84

COMPOUND : **Chlorothiophos**

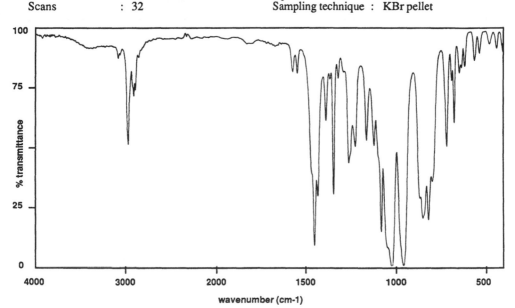

Chemical name   : O,O-diethyl O-[dichloro(methylthio)phenyl] phosphorothioate

Other name	: Celathion		
Type	: insecticide, acaricide		
Brutoformula	: C11H15Cl2O3PS2	CAS nr	: 60238-56-4
Molecular mass	: 361.24016	Exact mass	: 359.9577269
Instrument	: Bruker IFS-85	Optical resolution	: 2 cm-1
Scans	: 32	Sampling technique	: KBr pellet

Band maxima with relative intensity :

402	18	438	7	534	9	562	12	616	14
646	17	676	38	717	48	822	80	854	79
959	99	1023	100	1082	85	1120	48	1163	46
1224	48	1261	55	1320	19	1348	69	1390	37
1435	69	1455	90	1551	17	1577	16	2922	26
2983	47								

COMPOUND : **Chlortoluron**

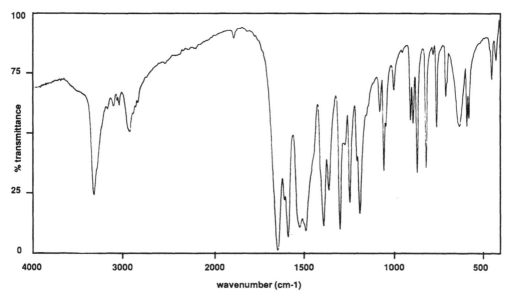

Chemical name   : 3-(3-chloro-p-tolyl)-1,1-dimethylurea

Other names   : Dicuran, Tolurex
Type   : herbicide
Brutoformula   : C10H13ClN2O   CAS nr   : 15545-48-9
Molecular mass   : 212.68091   Exact mass   : 212.0716334
Instrument   : Bruker IFS-85   Optical resolution : 2 cm-1
Scans   : 32   Sampling technique : KBr pellet

Band maxima with relative intensity :

402	14	422	20	445	28	571	44	583	48
625	48	703	35	756	48	816	65	867	67
889	46	903	45	995	32	1050	66	1072	41
1188	84	1244	79	1298	91	1358	74	1388	89
1487	91	1587	94	1647	100	2921	49	3318	75

COMPOUND : **Clofentezine**

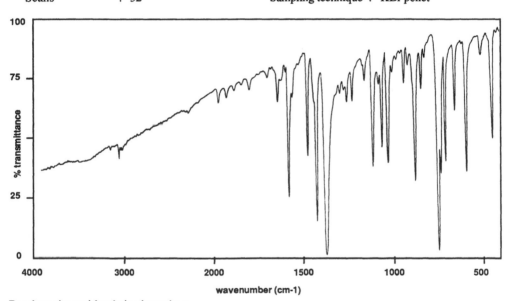

Chemical name   : 3,6-bis (2-chlorophenyl)-1,2,4,5-tetrazine

Other name	: Apollo, Bisclofentezin		
Type	: acaricide		
Brutoformula	: C14H8Cl2N4	CAS nr	: 74115-24-5
Molecular mass	: 303.15266	Exact mass	: 302.0125958
Instrument	: Bruker IFS-85	Optical resolution	: 2 cm-1
Scans	: 32	Sampling technique	: KBr pellet

Band maxima with relative intensity :

454	51	514	15	600	65	663	39	717	60
744	65	759	98	852	29	890	68	948	27
1039	60	1071	54	1123	62	1168	25	1237	34
1268	34	1386	100	1437	85	1486	57	1594	74
1654	34	1982	34	3086	57				

COMPOUND : **CME 134**

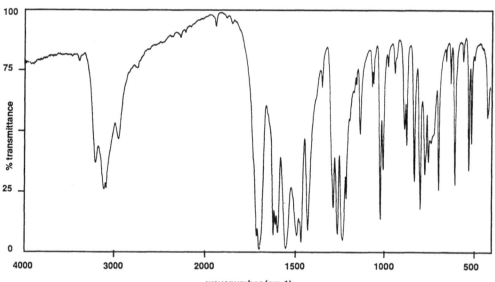

Chemical name	:	1-(3,5-dichloro-2,4-difluorophenyl)-3-(2,6-difluorobenzoyl) urea

Other name : Teflubenzuron
Type : acaricide, insecticide
Brutoformula : C14H6Cl2F4N2O2   CAS nr   : -
Molecular mass : 381.11572   Exact mass   : 379.9742372
Instrument : Bruker IFS-85   Optical resolution : 2 cm-1
Scans : 32   Sampling technique : KBr pellet

Band maxima with relative intensity :

420	44	508	55	522	66	551	20	598	72
620	29	691	74	747	62	768	68	795	82
828	71	870	55	882	49	936	25	1001	66
1018	87	1064	31	1133	51	1214	78	1235	96
1264	93	1287	82	1350	31	1432	91	1470	96
1494	94	1555	99	1600	92	1625	93	1704	100
2971	53	3137	74	3233	63				

88

COMPOUND : **Coumaphos**

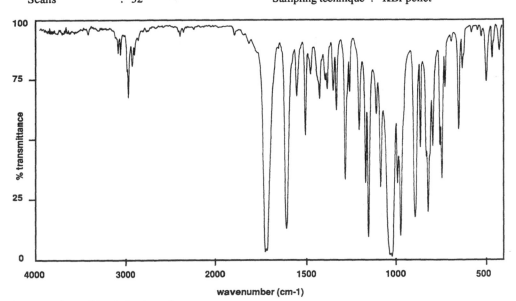

Chemical name : O-(3-chloro-4-methylcoumarine-7-yl) O,O-diethylphosphorothioate

Other names   : Cumafos, Co-ral, Resitox, Bayer 21199, Asuntol
Type   : insecticide
Brutoformula   : C14H16ClO5PS   CAS nr   : 56-72-4
Molecular mass : 362.76738   Exact mass   : 362.0144539
Instrument   : Bruker IFS-85   Optical resolution : 2 cm-1
Scans   : 32   Sampling technique : KBr pellet

Band maxima with relative intensity :

423	12	461	15	493	25	627	19	647	45
722	27	740	66	750	52	788	52	815	80
860	53	889	83	968	91	986	68	1013	100
1080	70	1105	39	1149	91	1165	68	1202	46
1257	29	1281	67	1331	37	1349	29	1382	28
1426	32	1476	22	1505	48	1555	31	1611	88
1729	98	2937	18	2979	32	3068	14	3096	13

COMPOUND : **Crimidine**

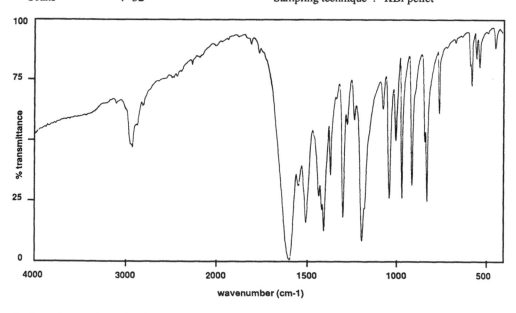

Chemical name   : 2-chloro-4-dimethylamino-6-methylpyrimidine

Other name	: Castrix		
Type	: rodenticide		
Brutoformula	: C7H10ClN3	CAS nr	: 535-89-7
Molecular mass	: 171.63085	Exact mass	: 171.0563187
Instrument	: Bruker IFS-85	Optical resolution	: 2 cm-1
Scans	: 32	Sampling technique	: KBr pellet

Band maxima with relative intensity :

442	10	531	18	548	15	574	26	755	38
826	75	909	68	965	74	997	49	1034	74
1070	36	1193	92	1233	40	1298	82	1368	64
1406	87	1503	84	1596	100	2924	52		

COMPOUND : **Crufomate**

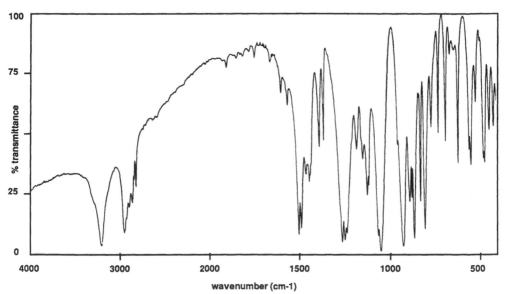

Chemical name : 4-tert.butyl-2-chlorophenyl-O-methyl-N-methyl phosphoroamidate

Other name : Ruelene, Dowco 152
Type : insecticide
Brutoformula : C12H19ClNO3P   CAS nr : 299-86-5
Molecular mass : 291.71689   Exact mass : 291.0791
Instrument : Bruker IFS-85   Optical resolution : 2 cm-1
Scans : 32   Sampling technique : KBr pellet

Band maxima with relative intensity :

419	47	442	48	468	62	516	37	544	63
554	57	615	62	660	16	686	53	727	49
766	47	804	90	828	78	865	94	876	77
889	79	924	97	1048	100	1123	76	1148	60
1182	56	1247	95	1263	96	1366	52	1389	55
1445	70	1488	89	1502	92	1564	37	1601	32
2830	71	2964	91	3219	96				

91

COMPOUND : **Cyanazine**

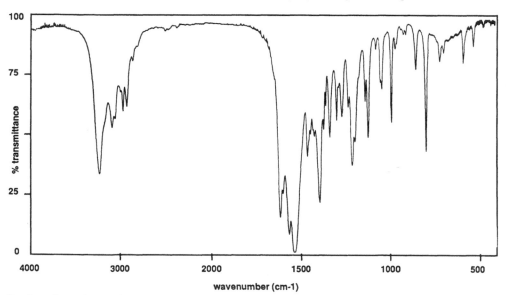

Chemical name   : 2-(4-chloro-6-ethylamino-1,3,5-triazin-2-ylamino)-2-methylpropionitrile

Other name   : Bladex, Blanchlol, Fortrol
Type   : herbicide
Brutoformula   : C9H13ClN6   CAS nr   : 21725-46-2
Molecular mass : 240.69716   Exact mass   : 240.0890121
Instrument   : Bruker IFS-85   Optical resolution   : 2 cm-1
Scans   : 32   Sampling technique : KBr pellet

Band maxima with relative intensity :

540	13	597	20	730	19	809	57	864	22
996	45	1052	30	1129	51	1145	36	1218	63
1275	42	1304	43	1343	51	1366	37	1402	78
1470	59	1547	100	1576	92	1626	85	2939	37
2984	39	3113	46	3256	66				

COMPOUND : **Cyanofenphos**

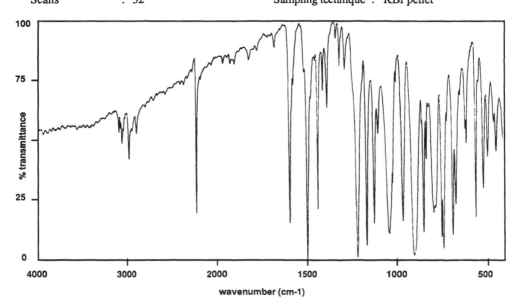

Chemical name   : O-ethyl-O-4-cyanophenyl-phenylphosphonothioate

Other names   : Surecide, SU 50
Type   : insecticide
Brutoformula   : C15H14NO2PS   CAS nr   :
Molecular mass   : 303.31809   Exact mass   : 303.0482825
Instrument   : Bruker IFS-85   Optical resolution : 2 cm-1
Scans   : 32   Sampling technique : KBr pellet

Band maxima with relative intensity :

447	55	495	57	519	71	562	83	614	51
674	77	689	90	740	96	751	91	794	81
835	58	851	89	904	99	964	84	1038	89
1102	47	1123	85	1165	94	1216	99	1284	19
1314	18	1333	6	1385	36	1410	28	1439	79
1499	100	1599	84	1682	10	1826	15	2231	80
2894	46	2982	56	3059	50	3095	45		

COMPOUND : **Cyclopropylcarboxylic acid-3-chloroanilide**

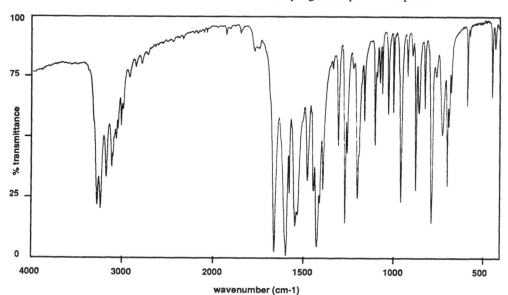

Chemical name	: cyclopropylcarboxylic acid-3-chloroanilide		

Other name	: -		
Type	: metabolite of Cyprofuram		
Brutoformula	: C10H10ClNO	CAS nr	: -
Molecular mass	: 195.650	Exact mass	: 195.045086
Instrument	: Bruker IFS-85	Optical resolution	: 2 cm-1
Scans	: 32	Sampling technique	: KBr pellet

Band maxima with relative intensity :

427	12	444	32	582	36	672	30	686	44
695	70	721	48	784	85	821	37	854	39
872	72	891	15	919	23	957	77	997	39
1026	39	1060	31	1075	26	1103	52	1162	42
1200	75	1258	56	1271	86	1308	53	1393	71
1426	96	1446	72	1478	67	1545	87	1578	73
1596	100	1661	98	3015	45	3122	62	3183	66
3246	80	3284	78						

COMPOUND : **Cycloxydim**

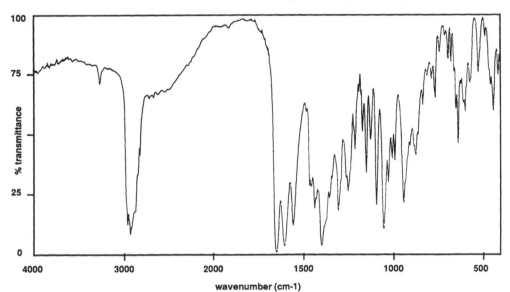

Chemical name   : 2-[1-(ethoxyimino)-butyl]-3-hydroxy-5-(2H-tetrahydrothiopyran-3-yl)-2-
                             cyclohexen-1-on

Other name	: Cycloxydim		
Type	: herbicide		
Brutoformula	: C17H27NO3S	CAS nr	:
Molecular mass	: 325.46964	Exact mass	: 325.1711524
Instrument	: Bruker IFS-85	Optical resolution	: 2 cm-1
Scans	: 32	Sampling technique	: KBr pellet

Band maxima with relative intensity :

411	24	435	39	520	23	568	27	594	40
632	53	645	39	675	18	690	17	738	14
760	34	781	26	831	37	870	58	937	78
985	60	1000	59	1021	69	1047	89	1088	79
1121	51	1145	65	1168	48	1209	56	1247	73
1301	82	1392	96	1433	81	1555	88	1601	97
1648	100	2930	92						

95

COMPOUND : **Cyhalothrin**

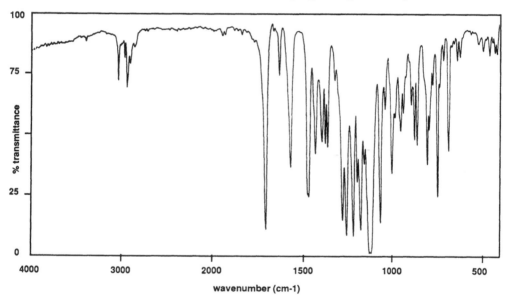

Chemical name	: R-α-cyano-3-phenoxybenzyl (1S)-cis-3-(Z-2-chloro-3,3,3-trifluoroprop-1-enyl)-2,2-dimethyl cyclopropanecarboxylate
Other name	: PP321
Type	: insecticide
Brutoformula	: C23H19ClF3NO3
Molecular mass	: 449.86098
Instrument	: Bruker IFS-85
Scans	: 32

CAS nr	:
Exact mass	: 449.1005426
Optical resolution	: 2 cm-1
Sampling technique	: KBr pellet

Band maxima with relative intensity :

434	15	465	16	500	14	628	16	644	18
695	55	722	16	757	76	802	47	813	62
869	54	884	51	901	32	946	40	960	48
1008	66	1049	38	1079	87	1134	100	1170	62
1191	90	1210	69	1233	92	1271	92	1294	86
1378	54	1391	53	1409	52	1446	57	1485	76
1585	63	1649	24	1724	89	2971	29	2997	16
3069	26								

96

COMPOUND : **Cyhexatin**

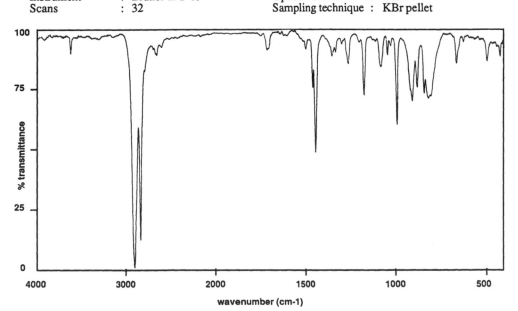

Chemical name   : tricyclohexyltin hydroxide

Other names   : Plictran, Dowco 213
Type   : acaricide
Brutoformula   : C18H34OSn   CAS nr   : 13121-70-5
Molecular mass   : 385.16108   Exact mass   : 386.1631505
Instrument   : Bruker IFS-85   Optical resolution : 2 cm-1
Scans   : 32   Sampling technique : KBr pellet

Band maxima with relative intensity :

418	11	490	13	660	14	818	29	841	27
879	24	906	30	990	40	1040	10	1078	15
1169	27	1257	14	1349	11	1444	51	1458	24
1715	8	2844	88	2915	100				

COMPOUND : **Cymiazole   (hydrochloride)**

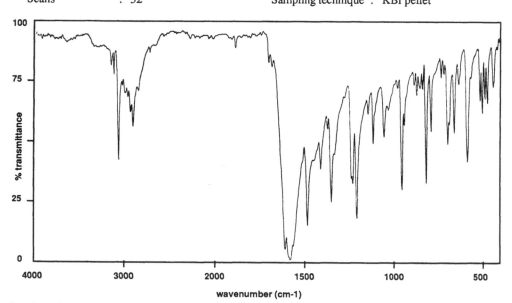

Chemical name	: 2-(2,4-dimethyl-phenylimino)-3-methyl-thiazolin (hydrochloride)

Other names	: CGA 192357, CGA 50439 hydrochloride		
Type	: insecticide		
Brutoformula	: C12H14N2S.HCl	CAS nr	: -
Molecular mass	: 254.77975	Exact mass	: 254.0644412
Instrument	: Bruker IFS-85	Optical resolution	: 2 cm-1
Scans	: 32	Sampling technique	: KBr pellet

Band maxima with relative intensity :

442	27	476	34	489	32	505	38	516	33
591	59	637	26	663	46	700	51	735	23
794	46	824	68	840	28	872	30	887	26
960	70	1060	48	1122	51	1149	38	1216	82
1237	67	1357	75	1414	61	1491	85	1588	100
1887	10	2913	42	3076	56	3125	20	3155	16

COMPOUND : **Cymoxanil**

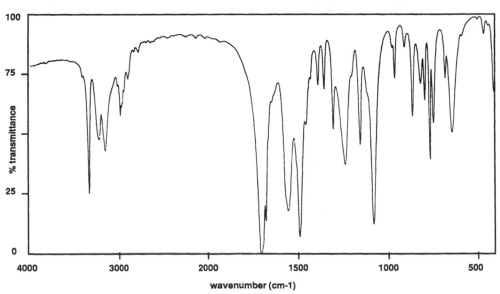

Chemical name	:	1-(2-cyano-2-methoxyiminoacetyl)-3-ethylurea		
Other names	:	Curzate, DPX 3217		
Type	:	fungicide		
Brutoformula	:	C7H10N4O3	CAS nr	: 57966-95-7
Molecular mass	:	198.18275	Exact mass	: 198.0752811
Instrument	:	Bruker IFS-85	Optical resolution	: 2 cm-1
Scans	:	32	Sampling technique	: KBr pellet

Band maxima with relative intensity :

458	7	637	48	673	25	740	44	760	60
788	35	811	27	857	42	902	12	958	25
1079	87	1152	53	1236	62	1302	47	1351	30
1384	28	1490	92	1555	81	1680	85	1708	100
2986	40	3158	55	3226	50	3335	73		

COMPOUND : **Cypermethrin**

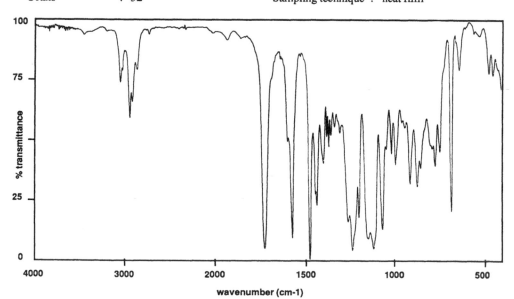

Chemical name : (RS)-α-cyano-3-phenoxybenzyl (1RS)-cis,trans-3-(2,2-dichlorovinyl)-
                  2,2-dimethylcyclopropanecarboxylate
Other names   : WL 43467, CGA 55186, CGA 109386
Type   : insecticide
Brutoformula   : C22H19Cl2NO3   CAS nr   : 52315-07-8
Molecular mass : 416.30763   Exact mass   : 415.0741891
Instrument   : Bruker IFS-85   Optical resolution : 2 cm-1
Scans   : 32   Sampling technique : neat film

Band maxima with relative intensity :

462	22	485	22	649	20	693	80	757	54
783	60	880	69	919	68	999	60	1023	55
1076	87	1126	95	1209	82	1246	96	1348	44
1368	47	1379	52	1391	48	1410	59	1448	77
1488	100	1586	90	1741	95	2958	39	3065	24

COMPOUND : **Cyproconazole**

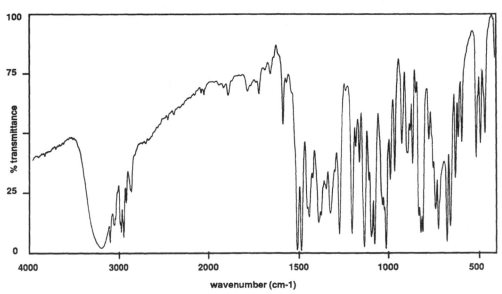

Chemical name : 1-N-(1,2,4-triazol)-2-(4-chlorophenyl)-3-cyclopropyl-2-butanol

Other names : SAN 619F
Type : fungicide
Brutoformula : C15H18N3OCl   CAS nr :
Molecular mass : 291.78321   Exact mass : 291.1138296
Instrument : Bruker IFS-85   Optical resolution : 2 cm-1
Scans : 32   Sampling technique : KBr pellet

Band maxima with relative intensity :

469	50	494	54	518	60	596	53	617	51
632	69	659	89	678	96	726	90	741	82
780	52	812	91	822	92	868	63	894	58
929	54	968	66	992	69	1013	99	1076	97
1095	94	1135	98	1163	62	1184	55	1205	92
1277	92	1328	84	1394	88	1448	85	1490	100
1515	99	1596	46	2876	74	2935	79	2963	94
2992	91	3119	96	3217	98				

COMPOUND : **Cyprofuram**

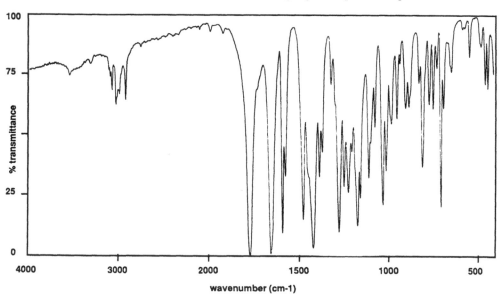

Chemical name   : N-(3-chlorophenyl)-N-tetrahydro-2-oxo-3-furanyl)cyclopropane-
                                    carboxamide
Other names   : Vinicur, SN 78314
Type   : fungicide
Brutoformula   : C14H14ClNO3   CAS nr   : 69581-33-5
Molecular mass   : 279.72558   Exact mass   : 279.066213
Instrument   : Bruker IFS-85   Optical resolution : 2 cm-1
Scans   : 32   Sampling technique : KBr pellet

Band maxima with relative intensity :

437	30	451	28	474	12	540	17	640	23
684	38	701	80	722	21	743	38	766	36
805	63	824	27	880	37	898	38	946	42
976	44	1006	64	1026	79	1072	46	1108	67
1156	76	1172	87	1223	73	1246	71	1275	90
1317	28	1367	56	1384	67	1420	97	1473	85
1572	66	1589	90	1654	99	1771	100	2907	33
3016	36	3061	29						

COMPOUND : **Cyromazin**

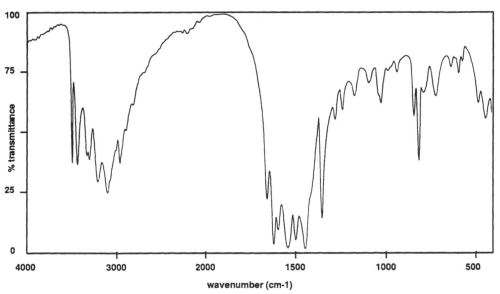

Chemical name   : 2-cyclopropylamino-4,6-diamino-1,3,5-triazine

Other names   : Trigard, CGA 72662, OMS-2014
Type   : insecticide
Brutoformula   : C6H10N6   CAS nr   : -
Molecular mass   : 166.1868   Exact mass   : 166.0966852
Instrument   : Bruker IFS-85   Optical resolution : 2 cm-1
Scans   : 32   Sampling technique : KBr pellet

Band maxima with relative intensity :

439	45	586	25	715	35	811	62	838	43
1024	38	1091	29	1170	35	1238	41	1355	87
1448	100	1500	96	1546	99	1625	97	1662	78
2970	62	3107	75	3213	70	3305	61	3438	63
3496	62								

COMPOUND : **2,4-D**

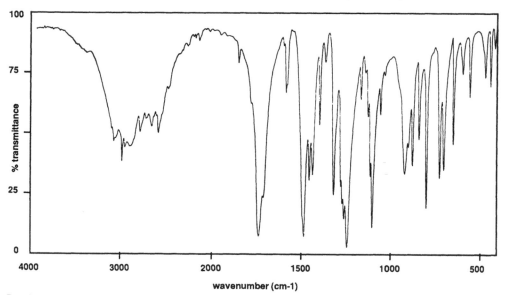

Chemical name	: (2,4-dichlorophenoxy)-acetic acid		
Other names	: Fernesta		
Type	: herbicide		
Brutoformula	: C8H6Cl2O3	CAS nr	: 94-75-7
Molecular mass	: 221.04122	Exact mass	: 219.9693961
Instrument	: Bruker IFS-85	Optical resolution	: 2 cm-1
Scans	: 32	Sampling technique	: KBr pellet

Band maxima with relative intensity :

415	15	438	31	468	27	554	36	595	26
645	56	697	67	720	70	794	83	837	54
871	65	914	68	1048	43	1093	91	1105	69
1119	44	1159	37	1232	100	1250	87	1310	77
1360	21	1392	48	1430	69	1448	71	1478	95
1585	34	1735	95	1852	22	2575	52	2649	49
2776	51	2977	64						

COMPOUND : **Dalapon-sodium**

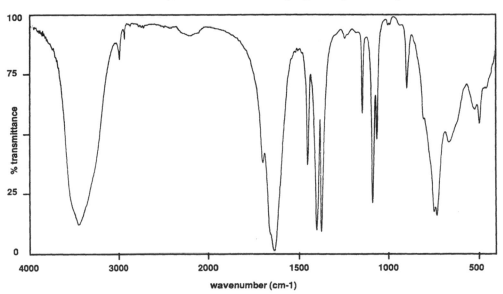

Chemical name   : sodium 2,2-dichloropropionate

Other names   : Dowpon, Killam
Type   : herbicide
Brutoformula   : C3H3Cl2O2Na   CAS nr   : 127-20-8
Molecular mass   : 164.95193   Exact mass   : 163.9407892
Instrument   : Bruker IFS-85   Optical resolution : 2 cm-1
Scans   : 32   Sampling technique : KBr pellet

Band maxima with relative intensity :

487	45	719	84	890	30	1055	52	1075	79
1135	41	1361	91	1388	91	1442	63	1625	100
2990	18	3440	89						

105

COMPOUND : **Daminozide**

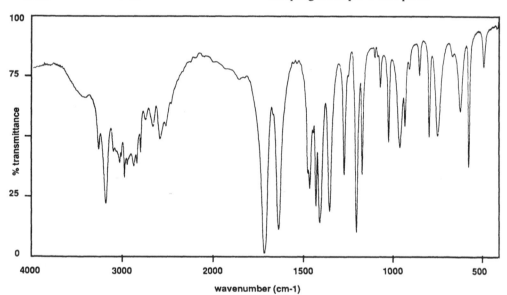

Chemical name	: N-dimethylaminosuccinamic acid		
Other names	: Alar, Kylar, B 995, B-nine		
Type	: growth regulator		
Brutoformula	: C6H12N2O3	CAS nr	: 1596-84-5
Molecular mass	: 160.17414	Exact mass	: 160.0847839
Instrument	: Bruker IFS-85	Optical resolution	: 2 cm-1
Scans	: 32	Sampling technique	: KBr pellet

Band maxima with relative intensity :

485	21	567	63	615	40	741	50	789	50
846	24	925	46	954	55	1015	53	1064	29
1166	66	1198	91	1267	66	1350	82	1405	86
1425	79	1460	72	1633	89	1711	100	2581	51
2659	46	2790	56	2866	62	2969	67	3024	61
3176	78	3257	55						

COMPOUND : **Dazomet**

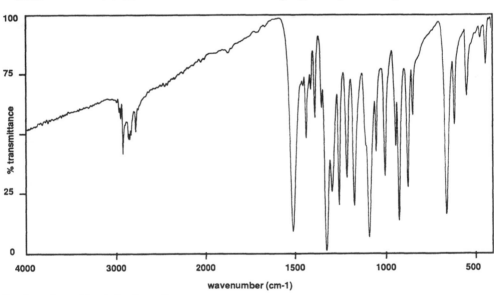

Chemical name	:	tetrahydro-3,5-dimethyl-1,3,5-thiadiazine-2-thione

Other names   : DMTT, Mylone, Stauffer N 521, Cragfungicide 974
Type   : herbicide, fungicide, nematicide

Brutoformula	: C5H10N2S2	CAS nr	:	533-74-4
Molecular mass	: 162.26885	Exact mass	:	162.0285378
Instrument	: Bruker IFS-85	Optical resolution	:	2 cm-1
Scans	: 32	Sampling technique	:	KBr pellet

Band maxima with relative intensity :

408	10	443	21	548	34	617	46	662	84
849	42	879	73	930	87	946	55	1007	68
1056	58	1096	94	1180	81	1220	69	1263	80
1301	75	1333	100	1357	39	1395	43	1417	31
1442	52	1516	91	2799	48	2866	52	2941	58
2967	43								

COMPOUND : **4,4'-DDT**

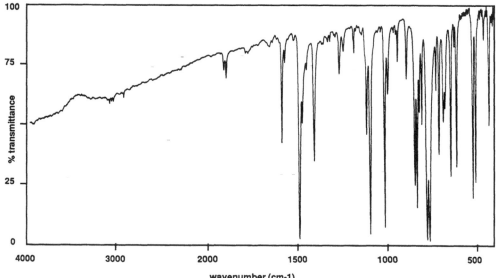

Chemical name : 1,1,1-trichloro-2,2-bis (4-chlorophenyl) ethane

Other names : Gesapon, Neocid, Didimac, Dedetane
Type : insecticide
Brutoformula : C14H9Cl5   CAS nr   : 50-29-3
Molecular mass : 354.49283   Exact mass   : 351.9146869
Instrument : Bruker IFS-85   Optical resolution : 2 cm-1
Scans : 32   Sampling technique : KBr pellet

Band maxima with relative intensity :

432	51	461	15	509	75	524	81	616	68
627	17	648	72	687	49	712	63	730	36
767	100	782	99	810	50	824	45	838	85
850	76	895	31	944	24	998	37	1015	94
1095	96	1114	54	1184	20	1243	19	1266	29
1407	65	1474	49	1492	98	1573	24	1590	58
1902	30								

108

COMPOUND : **2,4'-DDT**

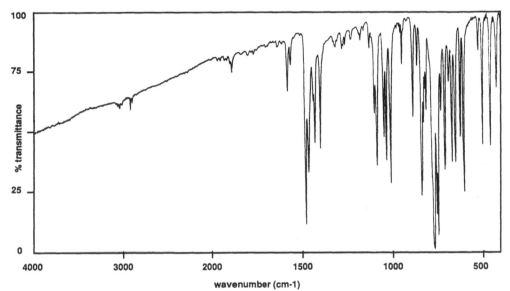

Chemical name   : 1,1,1-trichloro-2-(2-chlorophenyl)- 2-(4-chlorophenyl) ethane

Other names   : -
Type   : insecticide
Brutoformula   : C14H9Cl5   CAS nr   : -
Molecular mass   : 354.49283   Exact mass   : 351.9146869
Instrument   : Bruker IFS-85   Optical resolution : 2 cm-1
Scans   : 32   Sampling technique : KBr pellet

Band maxima with relative intensity :

427	31	461	56	505	55	526	16	608	75
626	52	653	63	673	63	689	29	713	66
734	41	753	94	759	85	774	100	815	40
831	46	843	77	866	22	890	43	951	21
1016	71	1039	62	1053	52	1093	64	1106	42
1133	14	1183	11	1286	14	1407	56	1436	54
1473	66	1489	88	1571	21	1590	32	1899	23
2937	39								

COMPOUND : **DDE**

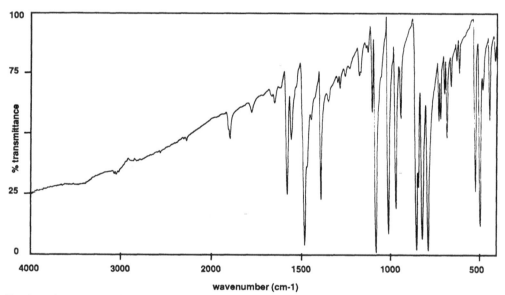

Chemical name : 1,1-dichloro-2,2-bis (4-chlorophenyl) ethene

Other names	: DDE, 4,4'-DDE		
Type	: insecticide		
Brutoformula	: C14H8Cl4	CAS nr	: 72-55-9
Molecular mass	: 318.0319	Exact mass	: 315.938009
Instrument	: Bruker IFS-85	Optical resolution	: 2 cm-1
Scans	: 32	Sampling technique	: KBr pellet

Band maxima with relative intensity :

411	19	443	44	500	88	525	74	612	24
626	18	658	29	683	51	696	32	719	43
729	44	794	99	827	94	860	98	943	43
973	81	1015	92	1088	100	1107	40	1178	24
1287	30	1396	77	1488	96	1562	51	1586	74
1653	36	1900	51						

COMPOUND : **Delta-keto-endrin**

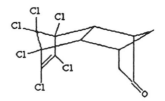

Chemical name	:	1,2,3,4,10,10-hexachloro-1,4,4a,5,6,7,8,8a-octahydro-6-keto-1,4:5,8-dimethanonaphtalene

Other names	: -			
Type	: insecticide			
Brutoformula	: C12H10Cl6O	CAS nr	: -	
Molecular mass	: 382.9309	Exact mass	: 379.8862787	
Instrument	: Bruker IFS-85	Optical resolution	: 2 cm-1	
Scans	: 32	Sampling technique	: KBr pellet	

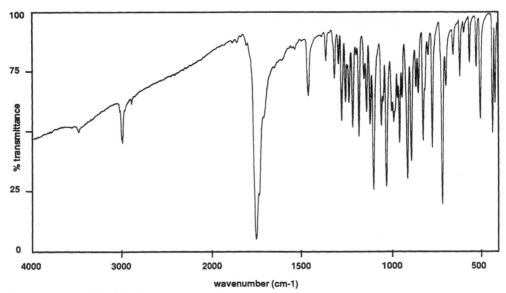

Band maxima with relative intensity :

416	41	430	54	497	48	520	26	558	24
612	30	649	21	689	34	710	85	767	60
790	21	817	57	844	37	858	35	885	66
907	74	939	39	953	58	964	39	985	50
1025	77	1052	51	1095	78	1114	50	1134	45
1148	31	1176	56	1211	52	1231	41	1251	41
1271	49	1289	25	1310	31	1358	24	1458	38
1751	100	2990	58						

111

COMPOUND   : **Deltamethrin**

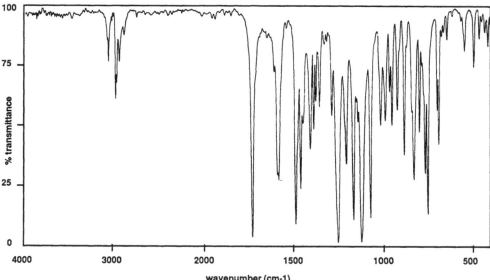

Chemical name   : (S)-a-cyano-3-phenoxybenzyl (1R)-cis-3-(2,2dibromovinyl)-2,2-
                    dimethylcyclopropanecarboxylate
Other names   : Decamethrin, Decis, Butox, K-othrin, RU 22974, NRDC 161
Type   : insecticide
Brutoformula   : C22H19Br2NO3   CAS nr   : 52918-63-5
Molecular mass   : 505.20963   Exact mass   : 502.9732629
Instrument   : Bruker IFS-85   Optical resolution   : 2 cm-1
Scans   : 32   Sampling technique   : KBr pellet

Band maxima with relative intensity :

416	17	464	14	494	26	546	19	645	14
690	58	700	44	753	88	768	71	801	53
832	73	885	63	923	44	953	50	966	36
989	48	1015	50	1072	89	1124	99	1167	90
1206	66	1252	100	1288	46	1355	42	1376	40
1387	52	1406	60	1460	77	1488	92	1584	73
1734	97	2924	22	2966	38	3048	22		

112

COMPOUND : **Demeton**

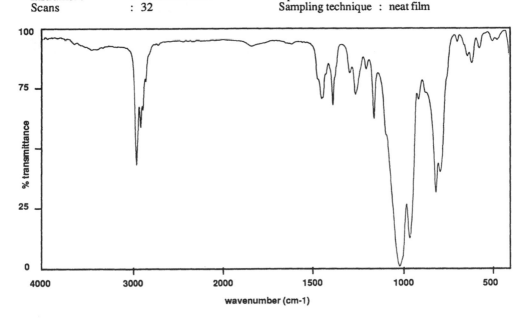

Chemical name	: O,O-diethyl O-[2-(ethylthio)] ethylphosphorothiate (mixture of 70% thionate and 30% thiolate)		
Other names	: Systemox		
Type	: insecticide		
Brutoformula	: C8H19O3PS2	CAS nr	: 8065-48-3
Molecular mass	: 258.33259	Exact mass	: 258.0513191
Instrument	: Bruker IFS-85	Optical resolution	: 2 cm-1
Scans	: 32	Sampling technique	: neat film

Band maxima with relative intensity :

611	13	819	68	970	88	1027	100	1163	36
1265	26	1390	31	1454	28	2929	39	2978	55

COMPOUND : **Demeton-S-methyl**

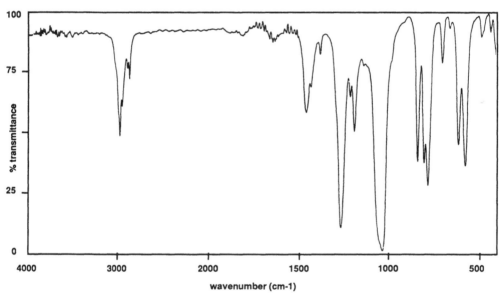

Chemical name : O,O-dimethyl S-[2-(ethylthio)ethylphosphorothiate

Other names : Metasystox, E 154, Duratox
Type : acaricide, insecticide
Brutoformula : C6H15O3PS2   CAS nr   : 919-86-8
Molecular mass : 230.27841   Exact mass   : 230.0200207
Instrument : Bruker IFS-85   Optical resolution : 2 cm-1
Scans : 32   Sampling technique : neat film

Band maxima with relative intensity :

481	9	566	64	602	55	692	20	772	72
793	62	828	62	1020	100	1182	49	1257	90
1377	17	1452	41	1641	12	2849	27	2952	51

114

COMPOUND : **Demeton-S-methyl-sulphone**

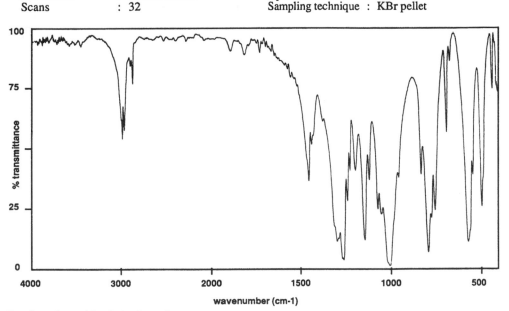

Chemical name : S-2-ethylsulphonylethyl O,O-dimethyl phosphorothioate

Other names : Metaisosystoxsulfon, E 158, Phosulfon
Type : insecticide
Brutoformula : C6H15O5PS2   CAS nr   : 17040-19-6
Molecular mass : 262.27721   Exact mass   : 262.0098489
Instrument : Bruker IFS-85   Optical resolution : 2 cm-1
Scans : 32   Sampling technique : KBr pellet

Band maxima with relative intensity :

426	25	491	75	571	90	663	13	685	43
753	76	793	94	830	61	1003	100	1067	76
1113	63	1140	88	1190	59	1219	59	1236	72
1260	97	1450	63	1723	8	1811	9	2856	21
2957	41	2977	44						

COMPOUND   : **Desethyl-atrazine**

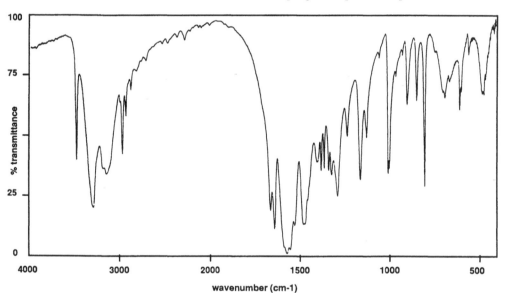

Chemical name   : 2-amino-4-isopropylamino-6-chloro-1,3,5-triazine

Other names   :
Type   : metabolite of Atrazine
Brutoformula   : C6H10CLN5
Molecular mass   : 187.6331
Instrument   : Bruker IFS-85
Scans   : 32

CAS nr   :
Exact mass   : 187.062465
Optical resolution   : 2 cm-1
Sampling technique : KBr pellet

Band maxima with relative intensity :

475	33	558	16	607	39	688	34	802	71
846	35	899	37	1004	66	1130	51	1165	68
1237	50	1292	75	1324	66	1340	64	1366	63
1383	64	1405	61	1482	87	1574	100	1642	89
1664	81	2932	41	2971	57	3151	65	3292	80
3469	59								

116

COMPOUND : **Desisopropyl-atrazine**

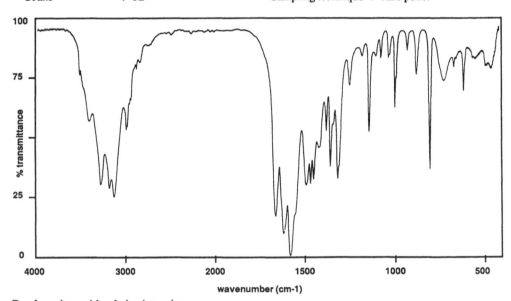

Chemical name   : 2-amino-4-ethylamino-6-chloro-1,3,5-triazine

Other names   :
Type   : metabolite of Atrazine
Brutoformula   : C5H8ClN5
Molecular mass   : 173.6060   CAS nr   :
Instrument   : Bruker IFS-85   Exact mass   : 173.046816
Scans   : 32   Optical resolution : 2 cm-1
                                       Sampling technique : KBr pellet

Band maxima with relative intensity :

457	20	611	29	722	25	796	62	876	22
927	12	994	36	1029	15	1072	15	1137	47
1244	27	1308	67	1350	62	1373	46	1444	67
1460	69	1483	70	1569	100	1610	90	1656	83
2985	46	3120	75	3267	70				

117

COMPOUND : **Desmedipham**

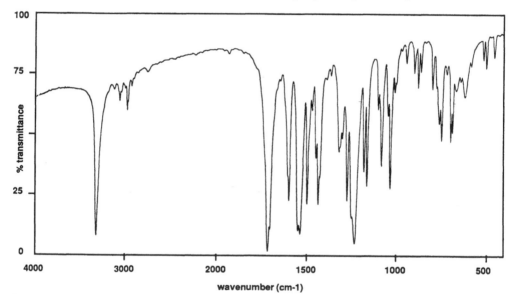

Chemical name : ethyl 3-phenylcarbamoyloxyphenylcarbamate

Other names : Betanal, SN 38107, Betanex
Type : herbicide
Brutoformula : C16H16N2O4   CAS nr : 13684-56-5
Molecular mass : 300.31692   Exact mass : 300.1109964
Instrument : Bruker IFS-85   Optical resolution : 2 cm-1
Scans : 32   Sampling technique : KBr pellet

Band maxima with relative intensity :

459	17	502	22	518	18	617	34	683	49
693	53	744	52	757	45	796	31	861	23
875	30	897	24	942	20	1003	32	1025	73
1038	42	1074	63	1094	39	1158	72	1174	66
1225	96	1269	78	1316	57	1432	79	1445	60
1494	79	1532	92	1596	78	1715	100	2978	40
3065	36	3324	93						

COMPOUND : **Desmethyl-pirimicarb**

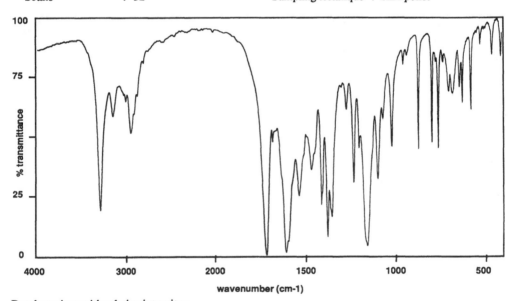

Chemical name : 5,6-dimethyl,-2-methylamino-4-pyrimidinyl dimethylcarbamate

Other names : metabolite of Pirimicarb
Type : herbicide
Brutoformula : C10H16N4O2   CAS nr   : 30614-22-3
Molecular mass : 224.2646   Exact mass   : 224.127315
Instrument : Bruker IFS-85   Optical resolution : 2 cm-1
Scans : 32   Sampling technique : KBr pellet

Band maxima with relative intensity :

454	15	518	10	569	38	618	35	633	28
671	31	693	30	751	54	785	52	862	55
948	19	1011	53	1090	67	1151	96	1195	54
1225	69	1266	38	1349	83	1373	92	1407	78
1463	63	1534	74	1607	98	1715	100	2945	47
3143	40	3284	80						

COMPOUND : **Desmetryne**

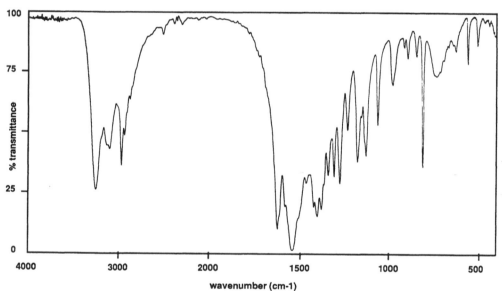

Chemical name : 2-isopropylamino-4-methylamino-6-methylthio-1,3,5-triazine

Other names	: Semeron, G 34360		
Type	: herbicide		
Brutoformula	: C8H15N5S	CAS nr	: 1014-69-3
Molecular mass	: 213.30225	Exact mass	: 213.1048077
Instrument	: Bruker IFS-85	Optical resolution	: 2 cm-1
Scans	: 32	Sampling technique	: KBr pellet

Band maxima with relative intensity :

503	13	556	21	730	26	807	64	841	17
889	18	973	30	1058	46	1125	59	1173	62
1230	49	1274	71	1306	68	1340	68	1402	85
1541	100	1623	90	2966	64	3103	57	3259	74

COMPOUND : **Dialifos**

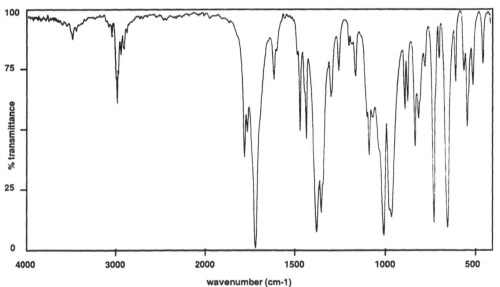

Chemical name	:	S-2-chloro-1-phtalimidoethyl O,O-diethyl phosphorodithioate

Other names	:	Torak, H 14503			
Type	:	acaricide, insecticide			
Brutoformula	:	C14H17ClNO4PS2	CAS nr	:	10311-84-9
Molecular mass	:	393.84265	Exact mass	:	393.0025103
Instrument	:	Bruker IFS-85	Optical resolution	:	2 cm-1
Scans	:	32	Sampling technique	:	KBr pellet

Band maxima with relative intensity :

448	22	504	31	536	49	553	25	598	30
649	91	689	20	724	89	768	23	806	45
828	57	867	38	885	41	962	87	1003	94
1082	61	1157	27	1249	25	1293	36	1351	85
1377	93	1431	53	1465	50	1612	28	1722	100
1765	50	1783	61	2897	15	2934	17	2979	38
3032	10								

COMPOUND : **Diallate**

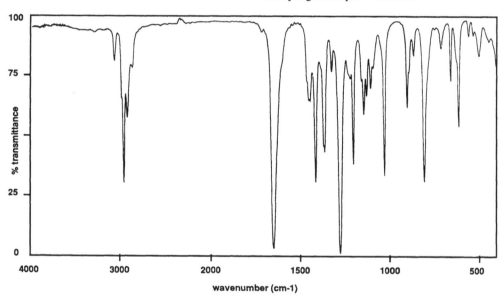

Chemical name   : S-2,3-dichloroallyl diisopropylthiocarbamate

Other names   : Avadex, CP 15336
Type   : herbicide
Brutoformula   : C10H17Cl2NOS   CAS nr   : 2303-16-4
Molecular mass   : 270.21909   Exact mass   : 269.0407844
Instrument   : Bruker IFS-85   Optical resolution   : 2 cm-1
Scans   : 128   Sampling technique   : neat film

Band maxima with relative intensity :

402	25	507	16	566	8	622	46	666	27
720	13	818	69	876	16	912	38	1036	67
1113	30	1136	33	1152	41	1210	62	1285	100
1332	22	1371	56	1422	69	1453	35	1660	97
2934	41	2974	69	3080	17				

COMPOUND   : **Diazinon**

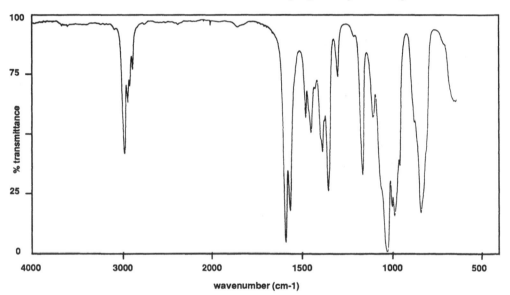

Chemical name	: O,O-diethyl O-(2-isopropyl-6-methylpirimidin-4-yl) phosphorothioate

Other names   : Basudine, Dimpylate, Sarolex, Neocidol, Diazol, Exodin
Type   : acaricide, insecticide, nematicide

Brutoformula	: C12H21N2O3PS	CAS nr	: 333-41-5
Molecular mass	: 304.34653	Exact mass	: 304.101042
Instrument	: Bruker IFS-85	Optical resolution	: 2 cm-1
Scans	: 32	Sampling technique	: KBr pellet

Band maxima with relative intensity :

832	83	981	84	1025	100	1099	43	1161	67
1294	25	1352	74	1381	57	1443	49	1471	42
1562	82	1587	95	2872	21	2931	35	2974	56

COMPOUND : **Dicamba**

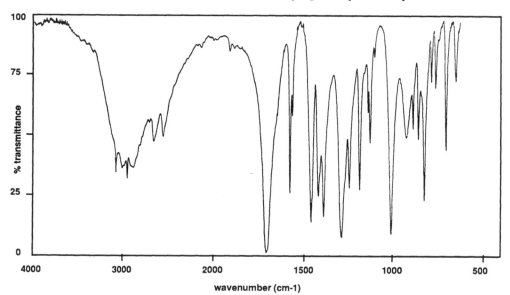

Chemical name   : 3,6-dichloro-o-anisic acid

Other names   : Banvel-D, Mediben, Banex
Type   : herbicide
Brutoformula   : C8H6Cl2O3   CAS nr   : 1918-00-9
Molecular mass   : 221.04122   Exact mass   : 219.9693961
Instrument   : Bruker IFS-85   Optical resolution   : 2 cm-1
Scans   : 32   Sampling technique   : KBr pellet

Band maxima with relative intensity :

646	27	702	56	757	30	782	27	823	78
855	52	884	47	919	51	1004	92	1124	53
1136	40	1184	73.	1242	72	1288	93	1389	84
1419	76	1460	87	1566	41	1580	74	1712	100
2550	50	2653	52	2948	68	3083	65		

124

COMPOUND : **Dichlobenil**

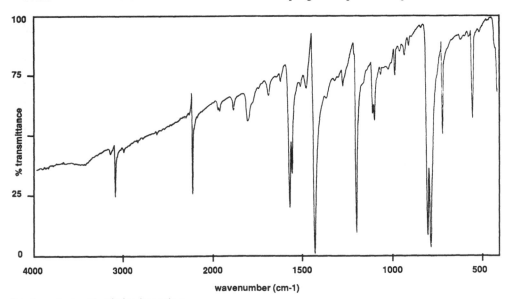

Chemical name	: 2,6-dichlorobenzonitrile		
Other names	: Casoron, H 133, WL 3379, Decabane, Norosac, Cyclomec		
Type	: herbicide		
Brutoformula	: C7H3Cl2N	CAS nr	: 1194-65-6
Molecular mass	: 172.01466	Exact mass	: 170.9642432
Instrument	: Bruker IFS-85	Optical resolution	: 2 cm-1
Scans	: 32	Sampling technique	: KBr pellet

Band maxima with relative intensity :

548	43	715	49	784	97	801	92	982	24
1098	43	1198	91	1270	28	1432	100	1476	29
1558	65	1571	80	1690	32	1807	43	1886	38
2231	74	3091	75						

COMPOUND : **Dichlofenthion**

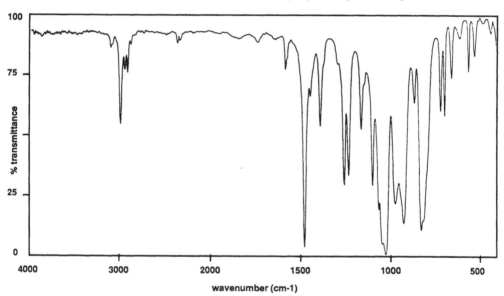

Chemical name   : O-(2,4-dichlorophenyl) O,O-diethyl phosphorothioate

Other names   : Nemacide, VC 13, Mobilawn
Type   : insecticide, nematicide
Brutoformula   : C10H13Cl2O3PS   CAS nr   : 97-17-6
Molecular mass   : 315.15307   Exact mass   : 313.970005
Instrument   : Bruker IFS-85   Optical resolution   : 2 cm-1
Scans   : 32   Sampling technique   : KBr pellet

Band maxima with relative intensity :

527	16	561	22	656	25	695	41	718	39
830	89	866	36	926	87	973	78	1024	100
1100	70	1163	47	1232	66	1256	70	1389	45
1477	96	1581	21	2905	22	2937	21	2983	44

126

COMPOUND : **Dichlofluanid**

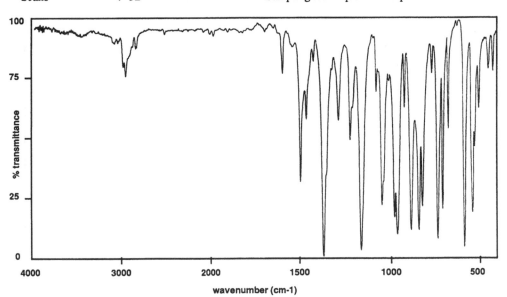

Chemical name : N-dichlorofluoromethylthio-N',N'-dimethyl-N-phenylsulphamide

Other names : Euparene, Elvaron, Bayer 47531
Type : acaricide, fungicide
Brutoformula : C9H11Cl2FN2O2S   CAS nr : 1085-98-9
Molecular mass : 333.22462   Exact mass : 331.962299
Instrument : Bruker IFS-85   Optical resolution : 2 cm-1
Scans : 32   Sampling technique : KBr pellet

Band maxima with relative intensity :

417	21	442	20	494	37	530	81	574	96
665	46	696	80	724	92	758	22	812	79
833	88	878	88	918	37	956	90	972	83
1040	78	1072	30	1157	97	1217	50	1280	42
1360	100	1455	41	1487	68	1588	22	2940	23

COMPOUND : **Dichloro-diphenyltin**

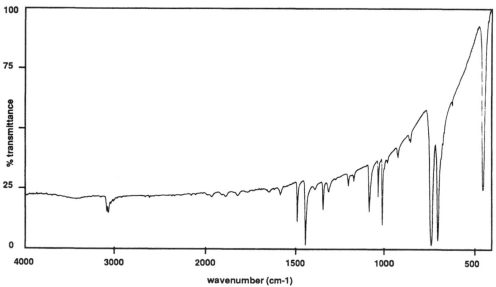

Chemical name : dichloro-diphenyltin

Other names : -
Type : fungicide
Brutoformula : C12H10Cl2Sn
Molecular mass : 343.809
Instrument : Bruker IFS-85
Scans : 32

CAS nr : -
Exact mass : 343.918152
Optical resolution : 2 cm-1
Sampling technique : KBr pellet

Band maxima with relative intensity :

442	76	691	97	729	100	996	91	1021	79
1073	85	1190	74	1332	84	1432	99	1480	90
3053	86								

COMPOUND : **4,4'-dichloro-benzophenon**

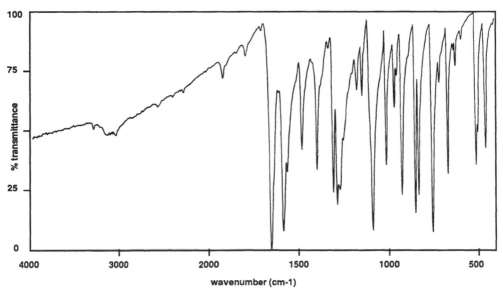

Chemical name   : 4,4'-dichloro-benzophenon

Other names   : -
Type   :
Brutoformula   : C13H8Cl2O   CAS nr   : 90-98-2
Molecular mass   : 251.1141   Exact mass   : 249.995217
Instrument   : Bruker IFS-85   Optical resolution : 2 cm-1
Scans   : 32   Sampling technique : KBr pellet

Band maxima with relative intensity :

458	57	510	64	623	22	667	68	715	29
754	93	831	77	851	85	927	77	967	40
1012	64	1087	92	1145	34	1173	32	1284	81
1306	76	1398	66	1482	57	1586	92	1653	100
1924	26	3035	51						

COMPOUND : **Dichloro-isocyanuric acid**

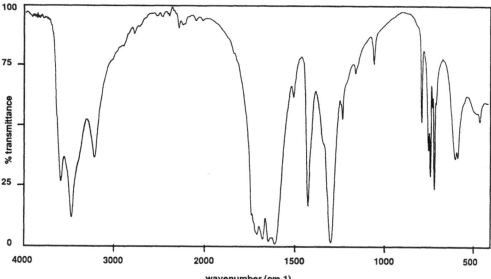

Chemical name : 1,3-dichloro-1,3,5-triazine-2,4,6-(1H,3H,5H) trione

Other names :
Type : herbicide
Brutoformula : C3HCl2N3O3   CAS nr   : 2782-57-2
Molecular mass : 155.94562   Exact mass   : 154.9302731
Instrument : Bruker IFS-85   Optical resolution : 2 cm-1
Scans : 32   Sampling technique : KBr pellet

Band maxima with relative intensity :

461	48	598	64	717	77	739	71	748	60
788	48	1056	24	1233	47	1301	99	1426	83
1611	100	1677	97	3222	63	3468	88	3588	73

COMPOUND : **1,2-Dichloropropane**

$$CH_3—CHCl—CH_2Cl$$

Chemical name   : 1,2-dichloropropane

Other names   :
Type   : insecticide, fungicide, nematicide
Brutoformula   : C3H6Cl2   CAS nr   : 78-87-5
Molecular mass   : 112.9873   Exact mass   : 111.984654
Instrument   : Bruker IFS-85   Optical resolution : 2 cm-1
Scans   : 32   Sampling technique : neat film

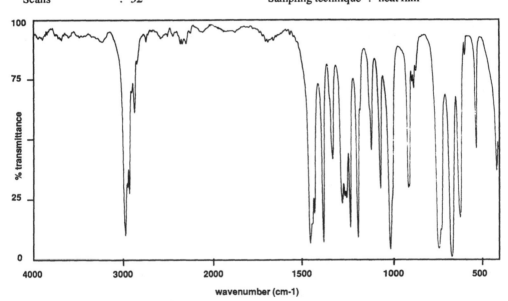

Band maxima with relative intensity :

414	63	527	54	620	83	670	100	741	96
878	28	914	70	1015	96	1066	70	1115	54
1191	91	1234	86	1253	74	1276	76	1326	57
1380	93	1429	80	1454	93	2870	37	2933	71
2982	89								

COMPOUND : **1,3-Dichloropropene** (cis)

$CHCl=CH-CH_2Cl$

Chemical name : 1,3-dichloropropene

Other names	: Telone, Vidden-D, Dipylene, DD
Type	: nematicide
Brutoformula	: C3H4Cl2
Molecular mass	: 110.97133
Instrument	: Bruker IFS-85
Scans	: 32

CAS nr	: 10061-02-6
Exact mass	: 109.9690046
Optical resolution	: 2 cm-1
Sampling technique	: neat film

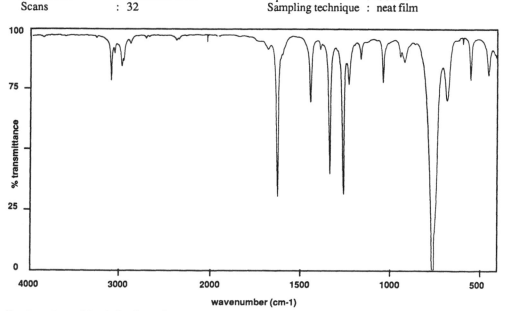

Band maxima with relative intensity :

446	17	547	19	679	28	766	100	914	12
1034	20	1158	10	1225	21	1258	67	1334	59
1440	28	1628	68	2965	13	3085	19		

COMPOUND : **1,3-Dichloropropene** **(trans)**

$$CHCl=CH-CH_2Cl$$

Chemical name : 1,3-Dichloropropene (trans)

Other names : -
Type : nematicide
Brutoformula : C3H4Cl2   CAS nr   : 10061-01-5
Molecular mass : 110.97133   Exact mass   : 109.9690046
Instrument : Bruker IFS-85   Optical resolution : 2 cm-1
Scans : 32   Sampling technique : neat film

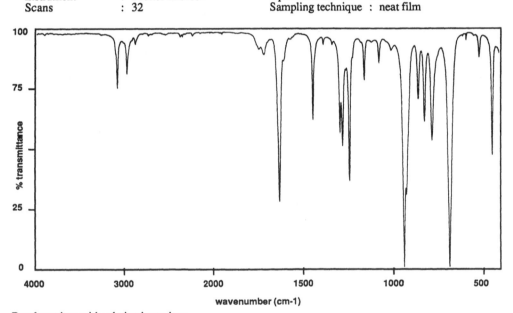

Band maxima with relative intensity :

446	52	516	11	682	100	780	46	821	38
857	28	936	100	1073	13	1155	20	1238	63
1276	48	1292	42	1442	37	1629	72	2960	17
3066	23								

COMPOUND : 2,6-Dichlorobenzamide

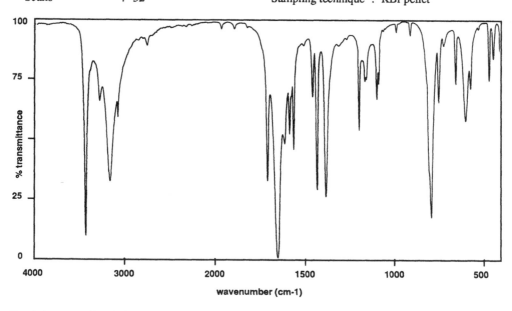

Chemical name	: 2,6-dichlorobenzamide		
Other names	: WL 3759, BAM		
Type	: metabolite of dichlobenyl		
Brutoformula	: C7H5Cl2NO	CAS nr	: 2008-58-4
Molecular mass	: 190.03	Exact mass	: 188.9748165
Instrument	: Bruker IFS-85	Optical resolution	: 2 cm-1
Scans	: 32	Sampling technique	: KBr pellet

Band maxima with relative intensity :

440	16	464	25	568	29	596	42	649	26
745	34	790	83	906	6	1096	33	1163	25
1197	46	1385	74	1433	71	1456	31	1562	54
1584	47	1650	100	1706	67	3077	39	3165	67
3432	90								

COMPOUND : **Dichlorprop**

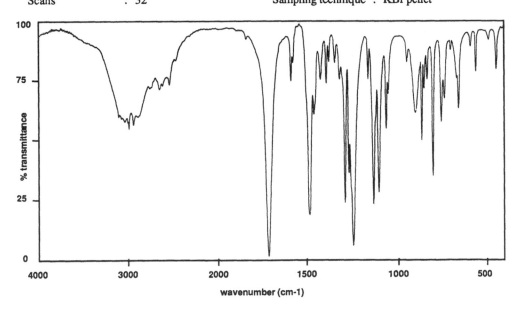

Chemical name   : 2-(2,4-dichlorophenoxy) propionic acid

Other names   : 2,4-DP
Type   : herbicide
Brutoformula   : C9H8Cl2O3   CAS nr   : 120-36-5
Molecular mass   : 235.06831   Exact mass   : 233.9850453
Instrument   : Bruker IFS-85   Optical resolution : 2 cm-1
Scans   : 32   Sampling technique : KBr pellet

Band maxima with relative intensity :

443	21	557	22	587	11	654	37	731	33
751	43	798	66	831	25	849	29	861	51
899	39	947	17	1061	46	1101	73	1130	78
1161	24	1242	95	1266	64	1287	77	1316	24
1342	17	1375	17	1389	26	1421	24	1457	39
1479	82	1587	25	1712	100	2940	43	2993	45

COMPOUND : **Dichlorvos**

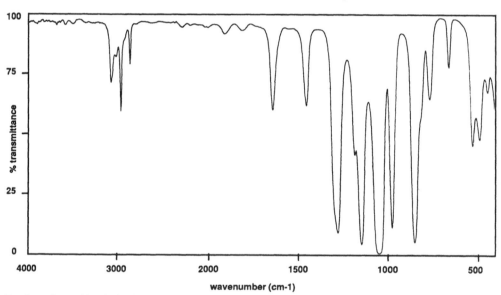

Chemical name   : 2,2-dichlorovinyl dimethyl phosphate

Other names	: DDVP, Dapona, UDVP		
Type	: acaricide, insecticide		
Brutoformula	: C4H7Cl2O4P	CAS nr	: 62-73-7
Molecular mass	: 220.97775	Exact mass	: 219.9458988
Instrument	: Bruker IFS-85	Optical resolution	: 2 cm-1
Scans	: 32	Sampling technique	: neat film

Band maxima with relative intensity :

491	52	530	54	659	21	769	34	856	95
979	89	1051	100	1151	96	1284	91	1456	37
1647	38	2860	18	2962	38	3072	26		

COMPOUND : **Diclofop-methyl**

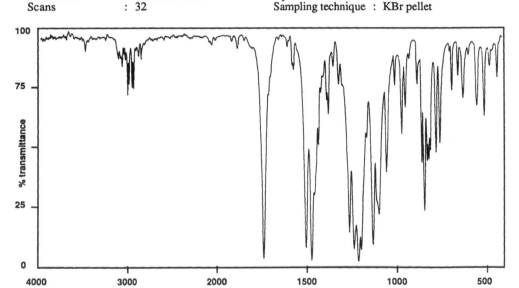

Chemical name   : methyl 2-[4-(2,4-dichlorophenoxy) phenoxy] propionate

Other names   : Illoxan, HOE 23408, Hoelon
Type   : herbicide
Brutoformula   : C16H14Cl2O4   CAS nr   : 51338-27-3
Molecular mass   : 341.19358   Exact mass   : 340.026907
Instrument   : Bruker IFS-85   Optical resolution : 2 cm-1
Scans   : 32   Sampling technique : KBr pellet

Band maxima with relative intensity :

437	22	478	17	509	38	549	34	626	30
654	21	690	27	757	49	779	53	820	56
829	57	846	78	860	58	885	24	949	34
971	45	1007	25	1056	62	1099	80	1134	92
1214	100	1239	94	1264	87	1321	24	1351	17
1379	37	1434	50	1474	99	1505	94	1572	18
1746	98	2842	13	2934	25	2984	23	2999	28
3059	16								

COMPOUND : **Dicloran**

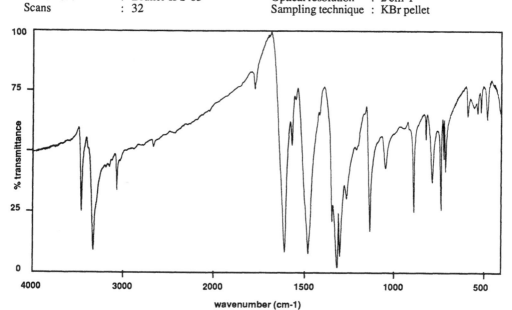

Chemical name   : 2,6-dichloro-4-nitro-aniline

Other names   : Allisan, Dichloran
Type   : fungicide
Brutoformula   : C6H4Cl2N2O2   CAS nr   : 99-30-9
Molecular mass   : 207.01698   Exact mass   : 205.9649792
Instrument   : Bruker IFS-85   Optical resolution : 2 cm-1
Scans   : 32   Sampling technique : KBr pellet

Band maxima with relative intensity :

482	37	516	34	535	35	589	36	714	59
723	54	741	75	789	64	823	46	894	76
1049	58	1144	84	1274	70	1315	95	1331	100
1358	80	1490	93	1574	48	1620	93	1776	24
3096	66	3358	91	3481	75				

COMPOUND : **Dicofol**

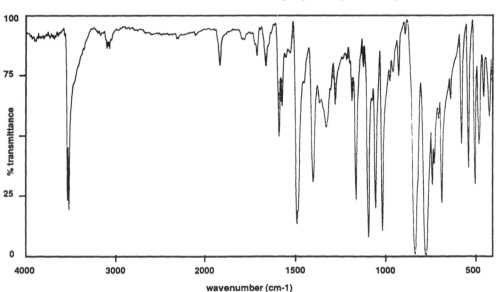

Chemical name   : 2,2,2-trichloro-1,1-bis (4-chlorophenyl) ethanol

Other names   : Kelthane, Acarin, FW 293, Hilfol, Mitigan, Cekudifol
Type   : acaricide
Brutoformula   : C14H9Cl5O   CAS nr   : 115-32-2
Molecular mass   : 370.49223   Exact mass   : 367.909601
Instrument   : Bruker IFS-85   Optical resolution : 2 cm-1
Scans   : 32   Sampling technique : KBr pellet

Band maxima with relative intensity :

416	41	447	32	474	53	497	69	533	63
570	52	629	33	679	77	720	61	731	70
767	100	830	99	919	23	1013	89	1053	79
1094	91	1118	19	1162	76	1183	34	1275	35
1325	45	1400	68	1490	86	1571	36	1589	49
1660	19	1711	14	1911	18	3525	79	3543	75

COMPOUND : **Dieldrin**

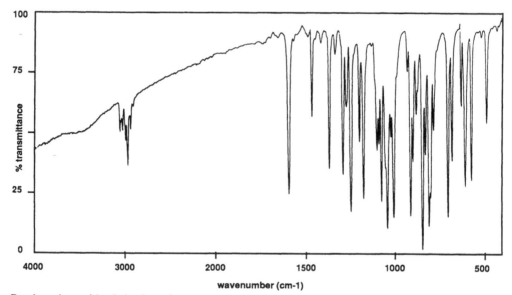

Chemical name	: (1R,4S,8R)-1,2,3,4,10,10-hexachloro-1,4,4a,5,6,7,8,8a-octahydro-6,7-epoxy-1,4:5,8-dimethanonaphtalene		
Other names	: HEOD		
Type	: insecticide		
Brutoformula	: C12H8Cl6O	CAS nr	: 60-57-1
Molecular mass	: 380.91496	Exact mass	: 377.8706295
Instrument	: Bruker IFS-85	Optical resolution	: 2 cm-1
Scans	: 32	Sampling technique	: KBr pellet

Band maxima with relative intensity :

487	46	571	70	606	73	628	39	679	62
703	86	783	49	809	90	830	59	845	100
879	42	898	62	911	85	929	24	1004	86
1018	52	1040	90	1075	79	1090	55	1102	57
1178	78	1200	54	1248	83	1273	39	1293	68
1336	17	1370	65	1416	12	1466	43	1596	76
2941	48	2973	63	2996	53	3061	49		

COMPOUND : **Dienochlor**

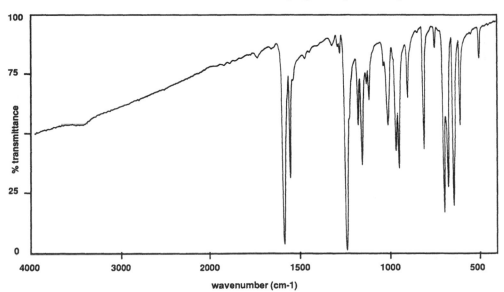

Chemical name   : 1,1',2,2',3,3',4,4',5,5'decachlorobi-2,4-cyclopentadien-1-yl

Other names   : Pentac
Type   : acaricide
Brutoformula   : C10Cl10   CAS nr   : 2227-17-0
Molecular mass   : 474.6415   Exact mass   : 469.688531
Instrument   : Bruker IFS-85   Optical resolution : 2 cm-1
Scans   : 32   Sampling technique : KBr pellet

Band maxima with relative intensity :

500	19	608	47	646	81	676	73	700	84
747	14	810	57	900	35	951	65	969	57
1012	47	1117	36	1160	63	1180	46	1248	100
1279	16	1558	68	1594	97				

COMPOUND   : **Diethathyl-ethyl**

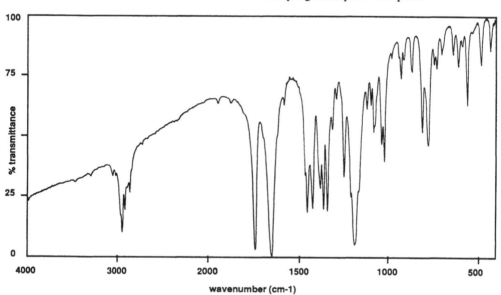

Chemical name   : N-(chloroacetyl)-N-(2,6-diethylphenyl) glycine ethyl ester

Other names   : Antor, H22234
Type   : herbicide
Brutoformula   : C16H22ClNO3   CAS nr   : 38727-55-8
Molecular mass   : 311.81164   Exact mass   : 311.1288098
Instrument   : Bruker IFS-85   Optical resolution : 2 cm-1
Scans   : 32   Sampling technique : KBr pellet

Band maxima with relative intensity :

432	14	483	19	562	37	607	20	636	15
701	15	728	21	779	53	811	48	865	22
926	25	1020	60	1033	53	1080	47	1095	36
1118	38	1196	95	1253	66	1317	46	1349	81
1370	80	1388	71	1432	80	1462	81	1664	100
1756	97	2934	80	2968	89				

142

COMPOUND : **Difenoxuron**

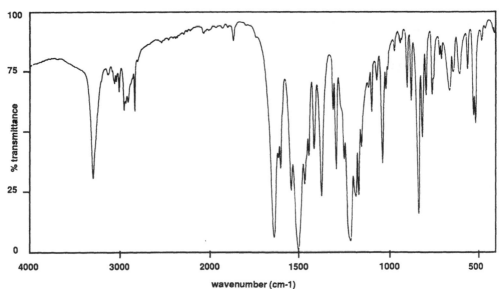

Chemical name : 3-[4-(4-methoxyphenoxy) phenyl]-1,1-dimethylurea

Other names	: C 3470, Lironion			
Type	: herbicide			
Brutoformula	: C18H18N2O3	CAS nr	: 14214-32-5	
Molecular mass	: 286.33346	Exact mass	: 286.1317315	
Instrument	: Bruker IFS-85	Optical resolution	: 2 cm-1	
Scans	: 32	Sampling technique	: KBr pellet	

Band maxima with relative intensity :

511	47	521	43	553	23	599	26	638	25
658	33	703	19	756	34	789	34	811	52
836	85	874	37	897	31	1014	32	1035	63
1065	28	1095	41	1153	56	1170	77	1186	77
1216	96	1248	61	1292	66	1308	40	1375	77
1413	57	1441	60	1465	72	1502	100	1542	74
1601	65	1641	94	1867	10	2834	40	2953	40
3005	32	3297	68						

COMPOUND : **Difenzoquat methyl sulphate**

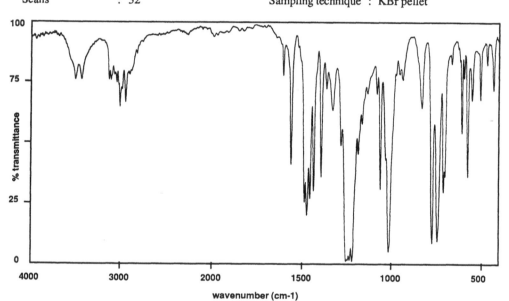

Chemical name   : 1,2-dimethyl-3,5-diphenylpyrazolium methyl sulphate

Other names	: AC 84777, Avenge, Finaven		
Type	: herbicide		
Brutoformula	: C18H20N2O4S	CAS nr	: 43222-48-6
Molecular mass	: 360.4311	Exact mass	: 360.1143675
Instrument	: Bruker IFS-85	Optical resolution	: 2 cm-1
Scans	: 32	Sampling technique	: KBr pellet

Band maxima with relative intensity :

433	27	468	16	505	31	552	31	577	64
596	22	609	45	710	70	743	91	773	92
830	35	936	23	1013	95	1059	69	1081	29
1186	54	1218	100	1253	99	1281	51	1329	36
1364	26	1394	64	1435	70	1457	73	1474	80
1488	74	1564	59	1608	21	2948	32	3014	34
3115	23								

COMPOUND : **Diflubenzuron**

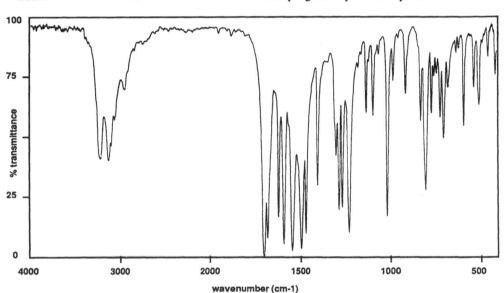

Chemical name	:	1-(4-chlorophenyl)-3-(2,6-difluorobenzoyl) urea		
Other names	:	Difluron, Dimilin, Astonex, PH-60-40, ENT 29054		
Type	:	insecticide		
Brutoformula	:	C14H9ClF2N2O2	CAS nr	: 35367-38-5
Molecular mass	:	310.68983	Exact mass	: 310.0320535
Instrument	:	Bruker IFS-85	Optical resolution	: 2 cm-1
Scans	:	32	Sampling technique	: KBr pellet

Band maxima with relative intensity :

418	23	458	15	509	36	537	28	594	44
635	13	681	28	704	50	724	41	744	22
771	39	803	72	830	42	916	31	983	25
1015	83	1096	40	1134	39	1226	90	1265	79
1282	80	1299	57	1403	70	1467	90	1494	96
1545	97	1593	94	1623	83	1684	92	1704	100
3142	59	3234	58						

COMPOUND : **Diflufenican**

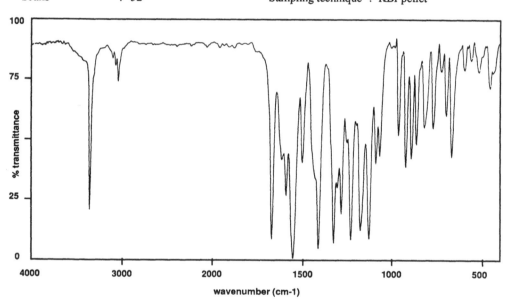

Chemical name	: N-(2,4-difluorophenyl)-2[(3-trifluoromethyl)-phenoxy]-3-pyridine-carboxamide
Other names	: M&B 38544
Type	: herbicide
Brutoformula	: C19H11F5N2O2
Molecular mass	: 394.30372
Instrument	: Bruker IFS-85
Scans	: 32

CAS nr	: -
Exact mass	: 394.0740562
Optical resolution	: 4 cm-1
Sampling technique	: KBr pellet

Band maxima with relative intensity :

460	27	520	20	563	16	599	20	665	56
696	39	725	20	767	44	817	44	862	51
889	57	919	61	962	47	1066	56	1091	59
1130	91	1178	87	1232	91	1286	80	1328	93
1413	95	1504	59	1554	100	1593	73	1672	91
3060	25	3371	79						

COMPOUND : **Dikegulac-sodium**

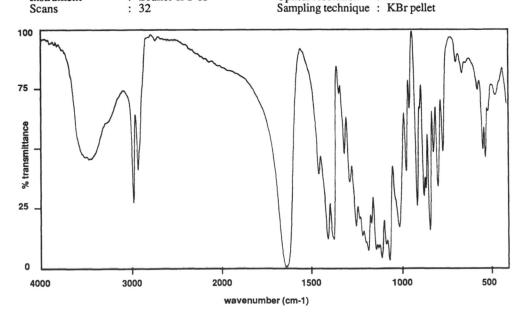

Chemical name   : sodium 2,3:4,6-di-O-isopropylidene-a-L-xylo-2-hexulofuranosonate

Other names   : Na-DAG, Atrinal, ACR-1032-C, RO 7-6145
Type   : growth regulator
Brutoformula   : C12H17O7Na   CAS nr   : 52508-35-7
Molecular mass   : 296.25486   Exact mass   : 296.0871979
Instrument   : Bruker IFS-85   Optical resolution : 2 cm-1
Scans   : 32   Sampling technique : KBr pellet

Band maxima with relative intensity :

529	53	542	50	657	17	763	51	791	66
815	51	838	84	871	69	913	74	953	32
973	59	1014	83	1068	97	1110	96	1187	92
1253	82	1286	63	1315	51	1375	88	1407	88
1457	60	1645	100	2939	57	2992	71	3477	53

COMPOUND   : **Dimefuron**

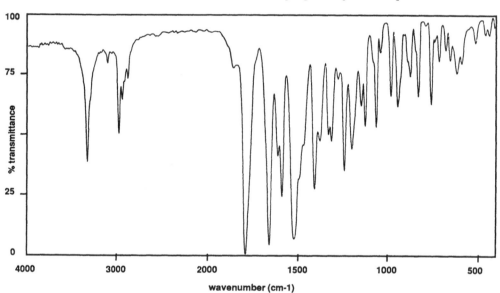

Chemical name   : 2-tert.butyl-4-[2-chloro-4-(3,3-dimethylureido)-phenyl]-5-oxo-1,3,4-
                   oxadiazoline
Other names   : -
Type   : herbicide
Brutoformula   : C15H19ClN4O3   CAS nr   :
Molecular mass   : 338.79668   Exact mass   : 338.1145556
Instrument   : Bruker IFS-85   Optical resolution : 2 cm-1
Scans   : 32   Sampling technique : KBr pellet

Band maxima with relative intensity :

433	7	509	10	611	23	648	18	671	13
707	18	752	36	823	33	867	24	935	37
973	33	1056	46	1118	45	1141	37	1193	55
1236	64	1307	52	1406	72	1521	93	1587	75
1660	96	1793	100	2974	48	3328	60		

COMPOUND : **Dimethachlor**

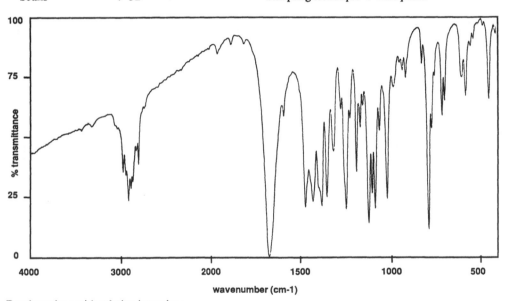

Chemical name	:	2-chloro-N-(2-methoxyethyl) acet-2'6'-xylidide

Other names	:	Teridox, CGA 17020			
Type	:	herbicide			
Brutoformula	:	C13H18ClNO2	CAS nr	:	50563-36-5
Molecular mass	:	255.74691	Exact mass	:	255.1025973
Instrument	:	Bruker IFS-85	Optical resolution	:	2 cm-1
Scans	:	32	Sampling technique	:	KBr pellet

Band maxima with relative intensity :

446	34	575	32	596	24	695	34	709	40
769	46	788	89	824	18	917	24	935	21
1022	75	1063	47	1088	79	1105	73	1124	86
1169	45	1190	64	1246	80	1314	55	1351	75
1380	78	1430	76	1471	79	1673	100	2810	60
2922	75	2983	63						

COMPOUND : **Dimethipin**

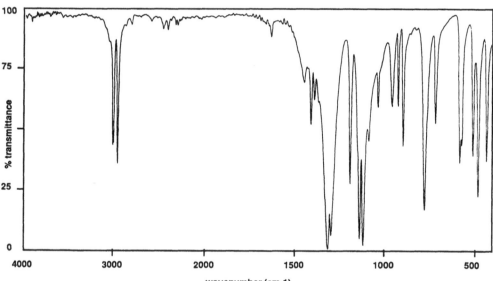

Chemical name : 2,3-dihydro-5,6-dimethyl-1,4-dithin-1,1,4,4-tetraoxide

Other names : UNI-N252
Type :
Brutoformula : C6H10S2O4   CAS nr   :
Molecular mass : 210.2642   Exact mass   : 210.0020478
Instrument : Bruker IFS-85   Optical resolution : 2 cm-1
Scans : 32   Sampling technique : KBr pellet

Band maxima with relative intensity :

430	62	476	77	505	60	576	63	709	46
770	83	890	56	917	39	950	39	1026	40
1109	98	1130	95	1183	72	1290	94	1308	100
1384	36	1404	47	1442	29	1627	10	2943	64
2995	56								

COMPOUND   : **Dimethoate**

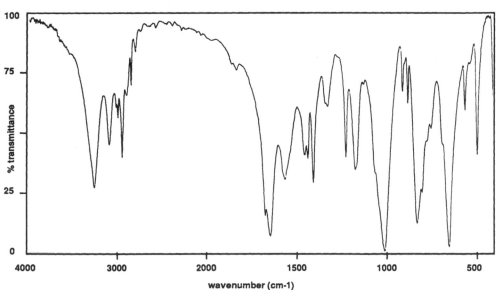

Chemical name   : O,O-dimethyl S-methylcarbamoylmethyl phosphorodithioate

Other names   : Rogor, Forthion, Demos, Perfekthion, Trimethion, Roxion
Type   : acaridide, insecticide
Brutoformula   : C5H12NO3PS2   CAS nr   : 60-51-5
Molecular mass   : 229.25005   Exact mass   : 228.9996201
Instrument   : Bruker IFS-85   Optical resolution : 2 cm-1
Scans   : 32   Sampling technique : KBr pellet

Band maxima with relative intensity :

406	35	495	59	562	40	651	98	831	88
881	37	909	32	1009	100	1173	65	1224	60
1327	38	1407	70	1437	60	1566	69	1650	93
2793	14	2841	28	2948	59	2993	42	3091	54
3260	72								

COMPOUND   : **Dinobuton**

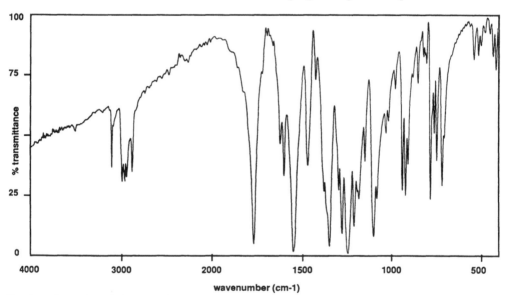

| Chemical name | : 2-sec-butyl-4,6-dinitrophenyl isopropyl carbonate |
</br>

Other names	: Icrex, Acrex, Dessin, MC 1053, Drawinol, Talan		
Type	: acaricide, fungicide		
Brutoformula	: C14H18N2O7	CAS nr	: 973-21-7
Molecular mass	: 326.30876	Exact mass	: 326.1113879
Instrument	: Bruker IFS-85	Optical resolution	: 2 cm-1
Scans	: 32	Sampling technique	: KBr pellet

Band maxima with relative intensity :

412	21	428	16	507	15	532	17	707	71
736	60	750	48	772	76	795	19	842	27
897	61	912	75	930	73	970	30	1024	48
1098	92	1145	60	1179	76	1207	88	1241	100
1273	91	1293	73	1345	96	1421	26	1463	62
1543	99	1596	67	1617	53	1760	95	2878	65
2938	67	2957	69	2989	69	3108	63		

152

COMPOUND : **Dinocap**

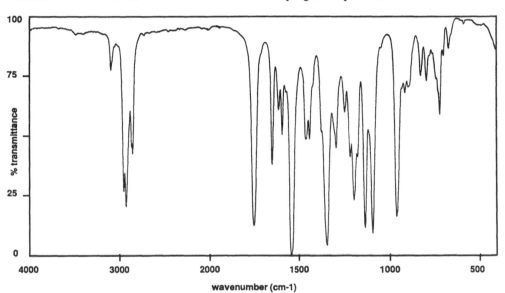

Chemical name	: 2 (or 4)-(1-methylheptyl)-4,6 (or2,6)-dinitrophenyl crotonate

Other names : Karathane, Ezenosan, Crotothane
Type : acaricide, fungicide

Brutoformula	: C18H24N2O6	CAS nr	: 39300-45-3
Molecular mass	: 364.40178	Exact mass	: 364.1634214
Instrument	: Bruker IFS-85	Optical resolution	: 2 cm-1
Scans	: 32	Sampling technique	: neat film

Band maxima with relative intensity :

674	12	723	40	795	26	827	24	961	84
1093	91	1135	88	1196	77	1217	58	1247	39
1296	54	1345	96	1441	51	1460	51	1544	100
1596	49	1618	38	1654	61	1756	87	2859	56
2930	79	2958	72	3104	21				

COMPOUND : **Dinoseb**

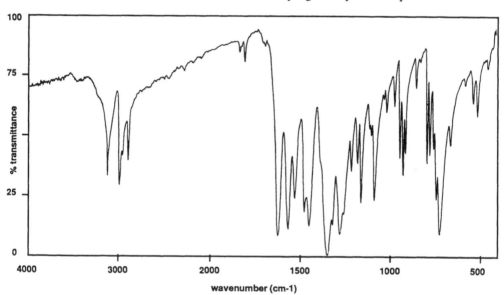

Chemical name   : 2-sec-butyl-4,6-dinitrophenol

Other names   : DNBP, Tribonate, Premerge, Gebutox, Basanite, Hivertox
Type   : herbicide, insecticide
Brutoformula   : C10H12N2O5   CAS nr   : 88-85-7
Molecular mass   : 240.21754   Exact mass   : 240.0746121
Instrument   : Bruker IFS-85   Optical resolution : 2 cm-1
Scans   : 32   Sampling technique : KBr pellet

Band maxima with relative intensity :

504	41	528	35	651	53	710	91	729	76
745	54	765	57	782	61	842	29	902	56
915	66	933	58	963	36	1005	39	1075	76
1150	77	1169	61	1204	64	1268	91	1337	100
1438	87	1466	81	1519	75	1555	88	1613	91
1801	18	2874	60	2970	70	3106	66		

COMPOUND : **Dinoseb-acetate**

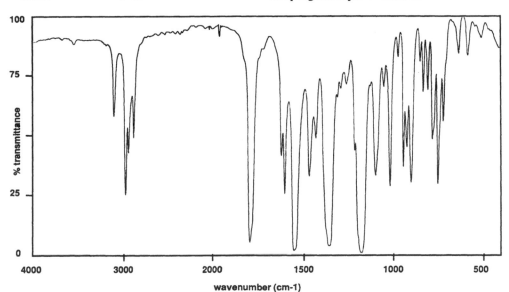

Chemical name	: 2-sec-butyl-4,6-dinitrophenyl acetate		

Other names	: Ivosit, Aretit, HOE 2764, Phenotan		
Type	: herbicide		
Brutoformula	: C12H14N2O6	CAS nr	: 2813-95-8
Molecular mass	: 282.25518	Exact mass	: 282.0851754
Instrument	: Bruker IFS-85	Optical resolution	: 2 cm-1
Scans	: 32	Sampling technique	: neat film

Band maxima with relative intensity :

507	8	580	16	631	15	713	44	742	70
772	51	798	30	823	31	840	18	890	70
915	55	934	63	967	16	1008	71	1044	29
1088	67	1167	100	1349	97	1424	51	1461	67
1546	99	1597	74	1617	58	1789	95	2877	51
2936	57	2968	75	3103	42				

COMPOUND : **Dinoterb**

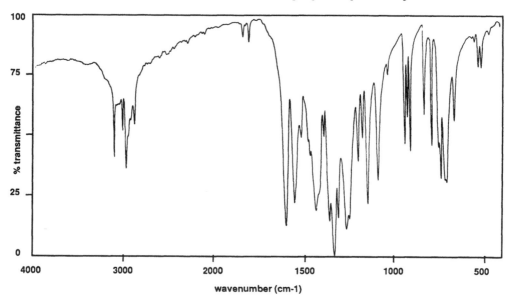

Chemical name	: 2-tert.-butyl-4,6-dinitrophenol		
Other names	: Herbogil, DNTBP		
Type	: herbicide		
Brutoformula	: C10H12N2O5	CAS nr	: 1420-07-1
Molecular mass	: 240.21754	Exact mass	: 240.0746121
Instrument	: Bruker IFS-85	Optical resolution	: 2 cm-1
Scans	: 32	Sampling technique	: KBr pellet

Band maxima with relative intensity :

510	21	526	21	659	43	704	69	734	67
784	53	824	40	902	56	918	41	933	53
1087	68	1148	78	1175	50	1201	60	1271	88
1316	84	1341	100	1367	85	1394	49	1441	81
1560	77	1607	87	1800	9	1832	7	2874	43
2972	62	3006	45	3106	57				

COMPOUND : **Dioxathion**

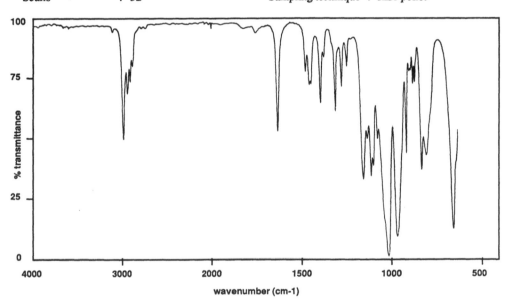

Chemical name : S,S'-(1,4-dioxane-2,3-diyl) O,O,O',O'-tetraethyl diphosphorodithioate

Other names : Delnav, AC 528, Deltic
Type : acaricide, insecticide
Brutoformula : C12H26O6P2S4   CAS nr   : 78-34-2
Molecular mass : 456.52494   Exact mass   : 456.008743
Instrument : Bruker IFS-85   Optical resolution : 2 cm-1
Scans : 32   Sampling technique : KBr pellet

Band maxima with relative intensity :

800	57	828	63	865	26	874	26	912	56
964	91	1013	100	1108	66	1152	67	1239	19
1270	27	1304	38	1389	34	1452	27	1474	21
1632	46	2901	24	2934	30	2981	49		

COMPOUND : **Biphenyl**

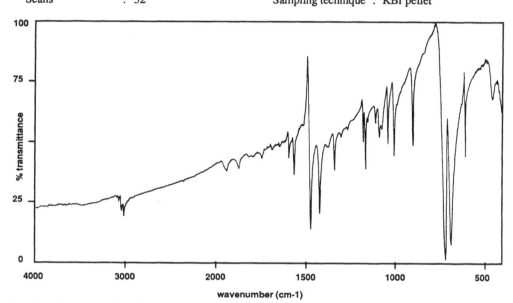

Chemical name	: 1-phenyl benzene		
Other names	: Diphenyl		
Type	: fungicide		
Brutoformula	: C12H10	CAS nr	: 92-52-4
Molecular mass	: 154.2135	Exact mass	: 154.078246
Instrument	: Bruker IFS-85	Optical resolution	: 2 cm-1
Scans	: 32	Sampling technique	: KBr pellet

Band maxima with relative intensity :

457	32	610	56	695	93	728	100	902	51
1005	55	1041	50	1090	47	1110	41	1169	61
1181	49	1344	61	1429	79	1479	86	1569	63
1597	55	3032	79						

COMPOUND : **Diphenylamine**

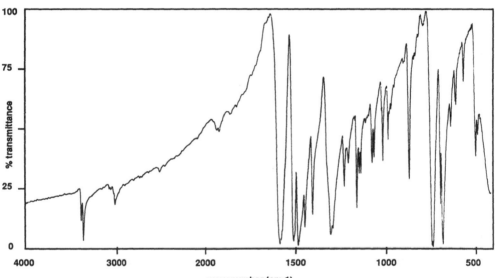

Chemical name	: N.N-diphenylamine			
Other names	: DPA			
Type	:			
Brutoformula	: C12H11N	CAS nr	: 122-39-4	
Molecular mass	: 169.22817	Exact mass	: 169.0891438	
Instrument	: Bruker IFS-85	Optical resolution	: 2 cm-1	
Scans	: 32	Sampling technique	: KBr pellet	

Band maxima with relative intensity :

493	52	505	61	568	30	612	39	642	49
689	98	701	81	743	100	875	71	993	54
1023	63	1073	61	1083	64	1148	68	1158	68
1172	83	1218	64	1242	73	1318	94	1417	85
1458	91	1494	98	1519	96	1596	98	1930	50
3039	80	3382	95	3406	87				

COMPOUND : **Diquat dibromide**

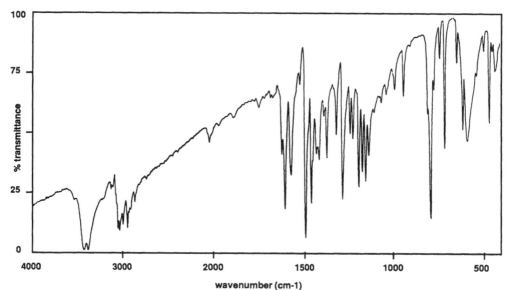

Chemical name : 1,1'-ethylene-2,2'-dipyridilium dibromide

Other names : Reglone, Reglox
Type : herbicide
Brutoformula : C12H12Br2N2   CAS nr : 85-00-7
Molecular mass : 344.05084   Exact mass : 341.9368216
Instrument : Bruker IFS-85   Optical resolution : 2 cm-1
Scans : 32   Sampling technique : KBr pellet

Band maxima with relative intensity :

436	24	467	46	497	16	587	54	612	49
645	21	712	57	739	19	791	87	939	35
1153	71	1171	67	1191	73	1224	53	1241	51
1285	78	1320	51	1373	61	1417	61	1462	80
1495	94	1576	68	1610	82	1627	59	2047	54
2949	90	3041	91	3388	100				

COMPOUND : **Disulfoton**

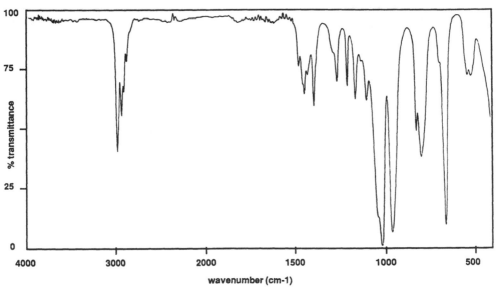

Chemical name	:	O,O-diethyl S-[2-(ethylthio) ethyl] phosphorodithioate

Other names	:	Disyston, Thiodemeton, Solvirex, Dithiosystox		
Type	:	acaricide, insecticide		
Brutoformula	:	C8H19O2PS3	CAS nr	: 298-04-4
Molecular mass	:	274.39319	Exact mass	: 274.0284777
Instrument	:	Bruker IFS-85	Optical resolution	: 2 cm-1
Scans	:	32	Sampling technique	: neat film

Band maxima with relative intensity :

404	52	516	27	658	91	797	62	825	51
959	94	1014	100	1097	37	1160	37	1204	31
1260	29	1388	40	1442	34	2928	43	2977	58

COMPOUND   : **Ditalimfos**

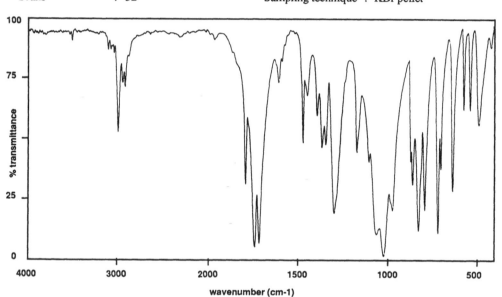

Chemical name   : O,O-diethyl phtalimidophosphonodithioate

Other names   : Millie, Dowco 199, Plondrel, Laptran, Frutogard
Type   : fungicide
Brutoformula   : C12H14NO4PS   CAS nr   : 5131-24-8
Molecular mass   : 299.28344   Exact mass   : 299.0381107
Instrument   : Bruker IFS-85   Optical resolution : 2 cm-1
Scans   : 32   Sampling technique : KBr pellet

Band maxima with relative intensity :

402	16	486	44	533	38	568	38	630	72
696	63	712	90	788	80	824	89	857	69
970	80	1018	100	1169	56	1295	81	1341	52
1363	54	1389	40	1443	32	1467	51	1603	26
1713	94	1738	95	1789	69	2908	28	2985	47

162

COMPOUND : **Dithianon**

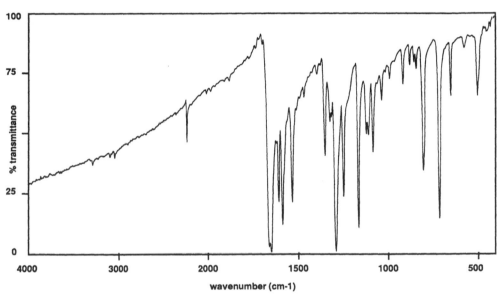

Chemical name   : 5,10-dihydro-5,10-dioxonaphto [2,3-b]-1,4-dithiin-2,3-dicarbonitrile

Other names   : Delan, IT931
Type   : fungicide
Brutoformula   : C14H4N2O2S2   CAS nr   : 3347-22-6
Molecular mass   : 296.32018   Exact mass   : 295.9714184
Instrument   : Bruker IFS-85   Optical resolution : 2 cm-1
Scans   : 32   Sampling technique : KBr pellet

Band maxima with relative intensity :

487	34	636	33	703	86	791	65	828	22
865	20	903	28	976	26	1022	35	1072	57
1098	50	1156	89	1239	76	1284	99	1313	44
1343	59	1526	78	1581	88	1601	78	1646	100
2224	52								

163

COMPOUND : **Diuron**

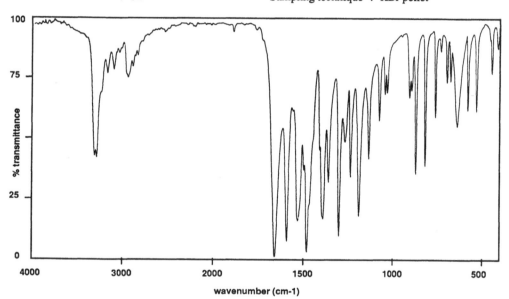

Chemical name   : 3-(3,4-dichlorophenyl)-1,1-dimethylurea

Other names   : Karmex, Diurex, Vonduron, Dynex, Unidron
Type   : herbicide
Brutoformula   : C9H10Cl2N2O   CAS nr   : 330-54-1
Molecular mass   : 233.09885   Exact mass   : 232.0170127
Instrument   : Bruker IFS-85   Optical resolution   : 2 cm-1
Scans   : 32   Sampling technique : KBr pellet

Band maxima with relative intensity :

410	11	442	22	526	38	574	37	632	44
671	25	690	25	724	12	756	40	813	61
865	64	900	32	1027	30	1039	31	1072	42
1133	58	1188	82	1235	66	1266	51	1299	91
1357	68	1387	83	1476	97	1525	84	1586	93
1653	100	2933	24	3091	20	3162	22	3281	58

COMPOUND : **DNOC**

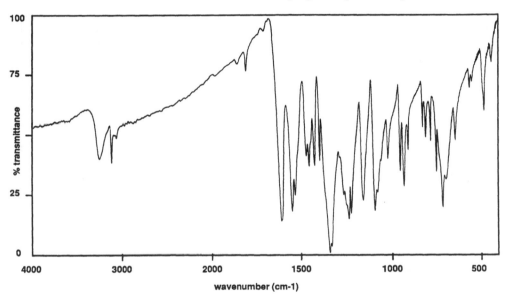

Chemical name : 4,6-dinitro-o-cresol

Other names : Dinitrocresol, DNC
Type : acaricide, insecticide, herbicide
Brutoformula : C7H6N2O5   CAS nr   : 534-52-1
Molecular mass : 198.13627   Exact mass   : 198.0276645
Instrument : Bruker IFS-85   Optical resolution : 2 cm-1
Scans : 32   Sampling technique : KBr pellet

Band maxima with relative intensity :

426	19	471	40	550	30	635	52	711	81
743	65	773	53	800	51	817	47	901	56
927	72	946	65	1014	60	1089	82	1152	77
1219	83	1232	85	1338	100	1390	60	1419	62
1449	63	1529	75	1547	82	1609	86	1799	22
3107	60	3239	58						

COMPOUND : **Dodemorph**

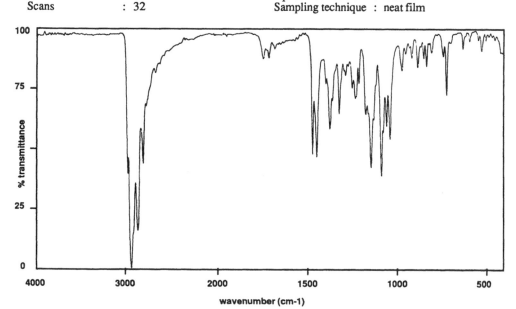

Chemical name : 4-cyclododecyl-2,5-dimethylmorpholine

Other names : Cyclomorph, Meltatox, Milban
Type : fungicide
Brutoformula : C18H35NO   CAS nr : 1593-77-7
Molecular mass : 281.48575   Exact mass : 281.2718483
Instrument : Bruker IFS-85   Optical resolution : 2 cm-1
Scans : 32   Sampling technique : neat film

Band maxima with relative intensity :

519	9	623	8	717	27	832	15	848	12
880	15	913	12	967	17	1033	46	1053	40
1083	61	1141	58	1207	22	1228	28	1319	35
1371	41	1444	53	1468	52	2802	55	2860	84
2930	100	2970	60						

166

COMPOUND　　:　**Dodine**

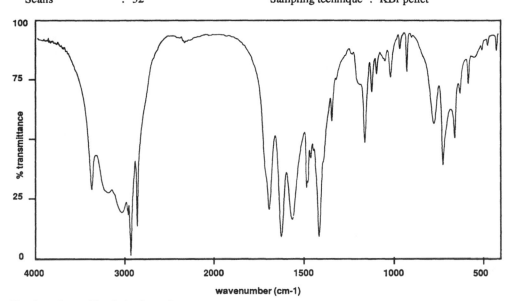

$C_{12}H_{25}$—NH—C—NH$_2$

NH

Chemical name	: 1-dodecylguanidinium acetate		
Other names	: Cyprex, Melprex, Carpene, Venturol, Syllit, Curitan		
Type	: fungicide		
Brutoformula	: C13H29N3	CAS nr	: 2439-10-3
Molecular mass	: 227.39618	Exact mass	: 227.236133
Instrument	: Bruker IFS-85	Optical resolution	: 2 cm-1
Scans	: 32	Sampling technique	: KBr pellet

Band maxima with relative intensity :

420	13	573	27	651	50	716	61	767	44
920	22	959	12	1009	24	1087	23	1114	30
1153	52	1339	43	1409	92	1478	71	1557	85
1618	92	1689	80	2851	87	2921	100	3369	72

COMPOUND   : **Flusilazol**

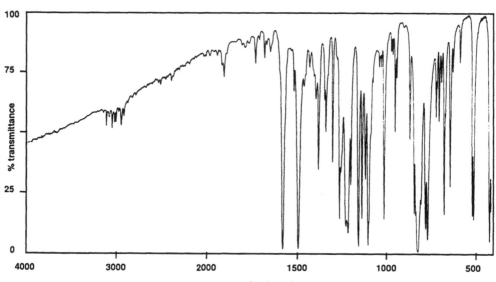

Chemical name   :

Other names   : DPX H6573
Type   : fungicide
Brutoformula   : C16H16N3F2Si   CAS nr   : 85509-19-9
Molecular mass   : 316.40832   Exact mass   : 316.1081466
Instrument   : Bruker IFS-85   Optical resolution   : 2 cm-1
Scans   : 32   Sampling technique   : KBr pellet

Band maxima with relative intensity :

419	82	428	95	514	86	521	84	589	19
628	23	643	72	676	84	692	28	705	44
723	42	771	94	783	90	827	100	846	83
871	52	946	26	956	49	1014	86	1044	21
1109	96	1124	69	1144	85	1163	97	1207	71
1222	91	1271	85	1310	61	1349	49	1390	65
1404	35	1439	21	1500	98	1525	31	1587	98
1738	20	1913	25	2962	46	3063	47	3129	46

168

COMPOUND   : **Endosulan (alpha)**

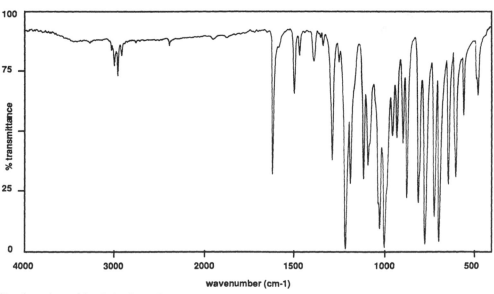

Chemical name   : 1,4,5,6,7,7-hexachloro-8,9,10-trinorbom-5-en-2,3-ylenedimethylsuplhite

Other names   : Thiodan, Beosit, HOE 2671, Cyclodan, Malix, Thiomul, Thionex
Type   : acaricide, insecticide
Brutoformula   : C9H6Cl6O3S   CAS nr   : 115-29-7
Molecular mass   : 406.92437   Exact mass   : 403.8168812
Instrument   : Bruker IFS-85   Optical resolution : 2 cm-1
Scans   : 32   Sampling technique : KBr pellet

Band maxima with relative intensity :

459	34	539	42	580	69	622	72	674	96
701	86	754	97	792	80	858	78	881	54
914	52	938	51	979	99	1004	91	1071	64
1096	70	1167	72	1192	100	1269	62	1372	20
1454	17	1482	33	1604	68	2936	26	2974	22

169

COMPOUND : **Endosulfan (beta)**

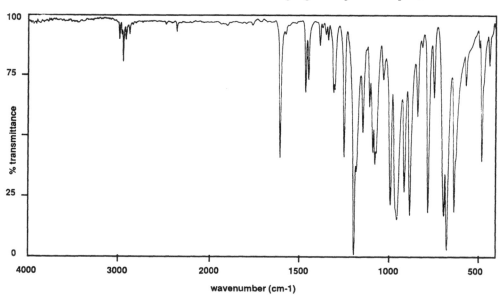

Chemical name   : 1,4,5,6,7,7-hexachloro-8,9,10-trinorborn-5-en-2,3-ylenedimethylsuplhite

Other names   : see Endosulfan (alpha)
Type   : acaricide, insecticide
Brutoformula   : C9H6Cl6O3S   CAS nr   : 115-29-7
Molecular mass   : 406.92437   Exact mass   : 403.8168812
Instrument   : Bruker IFS-85   Optical resolution : 2 cm-1
Scans   : 32   Sampling technique : KBr pellet

Band maxima with relative intensity :

436	19	479	60	568	28	633	81	675	97
691	83	744	33	779	8	838	41	881	83
912	73	953	85	990	78	1028	25	1076	61
1088	56	1108	37	1144	48	1193	100	1248	58
1308	31	1337	9	1384	11	1448	25	1464	31
1606	58	2923	9	2951	18	2992	8		

COMPOUND : **Endosulfan sulfate**

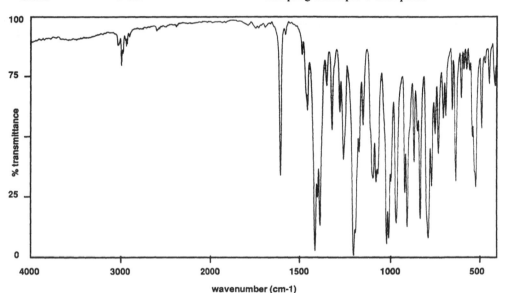

Chemical name : 1,4,5,6,7,7-hexachloro-8,9,10-trinorborn-5-en-2,3-ylenedimethylsulphate

Other names : Endosulfan sulfate
Type : metabolite of endosulfan
Brutoformula : C9H6Cl6O4S   CAS nr   : 1031-07-8
Molecular mass : 422.92377   Exact mass   : 419.8117953
Instrument : Bruker IFS-85   Optical resolution : 2 cm-1
Scans : 32   Sampling technique : KBr pellet

Band maxima with relative intensity :

406	28	439	27	482	46	516	71	565	21
580	21	596	33	628	68	648	38	684	40
699	41	726	57	744	48	764	70	784	92
828	84	863	60	902	87	916	73	964	86
1004	92	1016	95	1076	68	1092	67	1147	44
1168	56	1199	100	1254	59	1273	39	1316	46
1345	28	1381	87	1397	75	1409	97	1450	38
1479	14	1604	65	2989	19				

171

COMPOUND : **Endothal sodium**

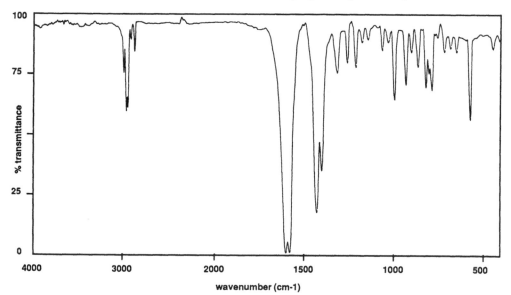

Chemical name : 7-oxabicyclo [2.2.1] heptane-2,3-dicarboxylic acid sodium salt

Other names : Prebetox, Aquathol
Type : herbicide
Brutoformula : C8H10O5Na2   CAS nr   : 129-97-0
Molecular mass : 232.16   Exact mass   :
Instrument : Bruker IFS-85   Optical resolution : 4 cm-1
Scans : 32   Sampling technique : KBr pellet

Band maxima with relative intensity :

572	44	646	15	680	14	717	15	788	31
821	30	865	21	900	15	931	29	993	35
1058	14	1172	11	1209	21	1257	19	1313	24
1404	65	1433	83	1588	100	1610	100	2871	14
2966	39	2993	23						

172

COMPOUND : **Endrin**

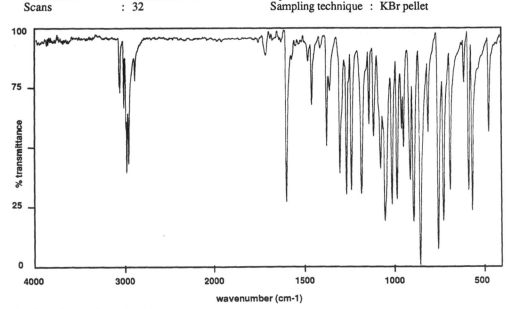

Chemical name   : 1,2,3,4,10,10-hexachloro-1,4,4a,5,6,7,8,8a-octahydro-6,7-epoxy-
                   1,4:5,8-dimethanonaphtalene
Other names   : Endrex, Hexadrin
Type   : insecticide, rodenticide
Brutoformula   : C12H8Cl6O   CAS nr   : 72-20-8
Molecular mass   : 380.91496   Exact mass   : 377.8706295
Instrument   : Bruker IFS-85   Optical resolution : 2 cm-1
Scans   : 32   Sampling technique : KBr pellet

Band maxima with relative intensity :

469	43	560	76	580	68	606	22	681	68
719	81	748	93	804	43	848	100	886	81
906	63	944	49	954	42	980	71	1007	74
1047	81	1073	58	1112	45	1137	40	1179	69
1235	68	1263	69	1299	60	1356	25	1374	49
1457	31	1478	13	1599	72	1714	10	2891	20
2960	56	2981	60	3014	32	3057	25		

173

COMPOUND : **EPTC**

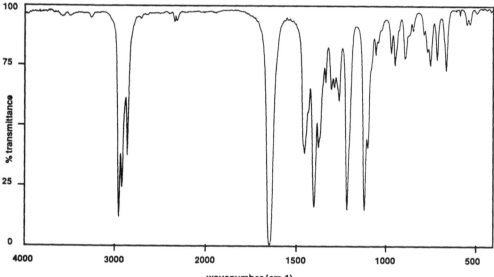

Chemical name   : S-ethyl dipropylthiocarbamate

Other names   : Eptam, R 1608
Type   : herbicide
Brutoformula   : C9H19NOS   CAS nr   : 759-94-4
Molecular mass   : 189.31788   Exact mass   : 189.1187274
Instrument   : Bruker IFS-85   Optical resolution : 4 cm-1
Scans   : 32   Sampling technique : neat film

Band maxima with relative intensity :

663	26	715	21	752	24	894	21	950	24
972	18	1056	19	1122	85	1220	85	1267	39
1342	31	1382	58	1406	83	1458	61	1650	100
1959	10	2875	62	2931	75	2964	88		

COMPOUND : **Ethidimuron**

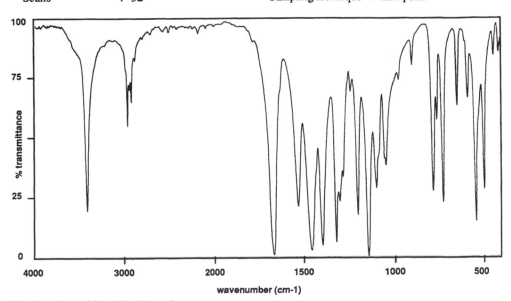

Chemical name : 1-(5-ethylsulphonyl-1,3-4-thiadiazol-2-yl)-1,3-dimethylurea

Other names : Sulfadiazol, Ustilan, Bayer 107033, Bayer 6579H
Type : herbicide
Brutoformula : C7H12N4O3S2   CAS nr   : 30043-49-3
Molecular mass : 264.31869   Exact mass   : 264.0350757
Instrument : Bruker IFS-85   Optical resolution : 2 cm-1
Scans : 32   Sampling technique : KBr pellet

Band maxima with relative intensity :

412	14	440	15	499	72	545	85	585	33
645	36	726	77	758	42	783	72	899	19
1048	61	1100	71	1147	100	1204	82	1240	30
1325	94	1402	95	1462	97	1534	78	1674	98
2924	33	2968	43	3417	79				

175

COMPOUND : **Ethiofencarb**

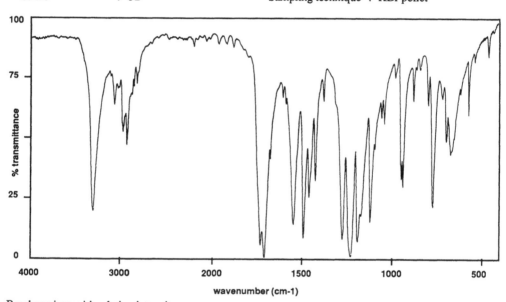

Chemical name : 2-[(ethylthio) methyl] phenyl methylcarbamate

Other names : Ethifencarb, Croneton, Bayer 108594, HOX 1901
Type : insecticide
Brutoformula : C11H15 NO2S
Molecular mass : 225.3077
Instrument : Bruker IFS-85
Scans : 32

CAS nr : 29973-13-5
Exact mass : 225.0823431
Optical resolution : 2 cm-1
Sampling technique : KBr pellet

Band maxima with relative intensity :

461	14	570	39	670	56	694	50	715	32
768	78	796	35	877	33	935	69	944	66
977	23	1036	43	1053	40	1116	84	1185	93
1223	99	1269	92	1375	33	1420	67	1456	74
1487	91	1542	85	1704	100	1727	94	2204	10
2817	26	2925	52	2967	47	3059	35	3293	80

COMPOUND : **Ethion**

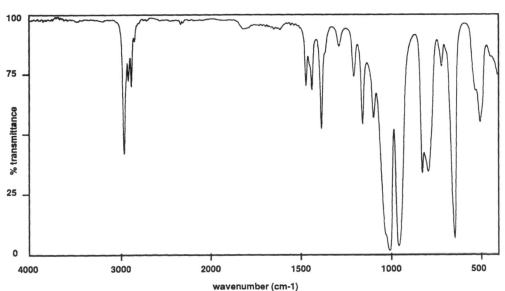

Chemical name : O,O,O',O'-tetraethyl S,S'-methylene di phosphorodithioate

Other names : Ethiol, Ethanox, Nialate, Dumuril,
Type : acaricide, insecticide
Brutoformula : C9H22O4P2S4
Molecular mass : 384.46081
Instrument : Bruker IFS-85
Scans : 32

CAS nr : 563-12-2
Exact mass : 383.9876164
Optical resolution : 2 cm-1
Sampling technique : neat film

Band maxima with relative intensity :

402	32	507	45	649	94	718	21	795	66
828	67	963	98	1012	100	1099	43	1160	46
1208	25	1290	12	1389	47	1441	31	1473	29
2901	29	2936	27	2982	58				

COMPOUND : **Ethirimol**

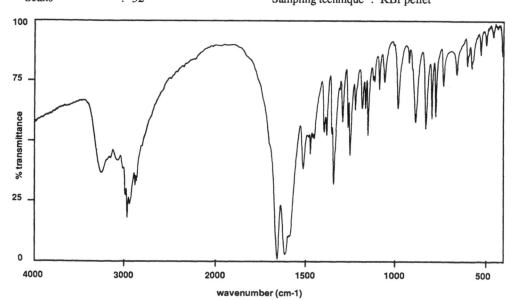

Chemical name	: 5-butyl-2-ethylamino-6-methylpirimidin-4-ol

Other names	: PP 149, Milgo, Milstem, Milcap		
Type	: fungicide		
Brutoformula	: C11H19N3O	CAS nr	: 23947-60-6
Molecular mass	: 209.29358	Exact mass	: 209.1528011
Instrument	: Bruker IFS-85	Optical resolution	: 2 cm-1
Scans	: 32	Sampling technique	: KBr pellet

Band maxima with relative intensity :

455	5	492	9	521	13	570	19	595	18
652	21	727	26	769	39	790	40	824	44
879	41	917	16	973	36	1046	25	1076	28
1104	24	1140	47	1155	36	1173	36	1212	36
1242	55	1254	44	1284	41	1333	68	1375	47
1389	46	1465	55	1505	61	1606	98	1649	100

COMPOUND   : **Ethofumesate**

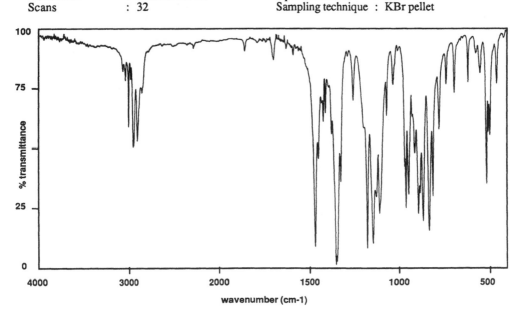

Chemical name	: 2-ethoxy-2,3-dihydro-3,3-dimethylbenzofuran-5-yl methanesulphonate		
Other names	: Nortran, NC 8438, Tramat, Progress		
Type	: herbicide		
Brutoformula	: C13H18O5S	CAS nr	: 26225-79-6
Molecular mass	: 286.34541	Exact mass	: 286.087486
Instrument	: Bruker IFS-85	Optical resolution	: 2 cm-1
Scans	: 32	Sampling technique	: KBr pellet

Band maxima with relative intensity :

469	23	508	45	516	43	527	65	560	18
628	22	704	27	749	23	787	42	821	71
842	85	876	81	903	78	923	52	957	70
972	76	1039		1076	36	1116	78	1154	91
1187	93	1265	30	1335	64	1362	100	1387	44
1421	35	1434	39	1461	54	1480	92	1715	12
2934	47	2982	49	3033	41	3069	21		

COMPOUND   : **Ethoprophos**

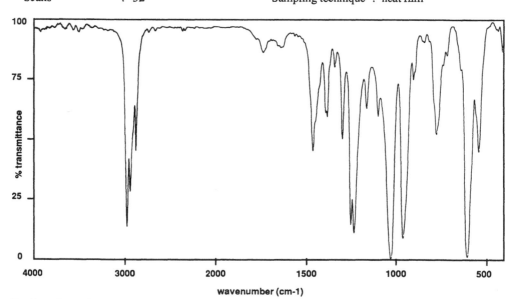

Chemical name   : O-ethyl S,S-dipropyl phosphorodithioate

Other names   : Ethoprop, Prophos, Mocap
Type   : insecticide, nematicide
Brutoformula   : C8H19O2PS2   CAS nr   : 13194-48-4
Molecular mass   : 242.33319   Exact mass   : 242.056405
Instrument   : Bruker IFS-85   Optical resolution : 2 cm-1
Scans   : 32   Sampling technique : neat film

Band maxima with relative intensity :

402	22	535	54	595	99	769	47	896	24
951	91	1017	100	1095	39	1159	36	1228	88
1247	85	1296	49	1338	18	1379	39	1458	54
2873	53	2933	71	2965	86				

180

COMPOUND : **Ethoxyquin**

Chemical name : 6-ethoxy-1,2-dihydro-2,2,4-trimethylquinoline

Other names : Santoquin, Nix-Scald
Type : growth regulator
Brutoformula : C14H19NO   CAS nr   : 91-53-2
Molecular mass : 217.31363   Exact mass   : 217.1466547
Instrument : Bruker IFS-85   Optical resolution : 2 cm-1
Scans : 32   Sampling technique : neat film

Band maxima with relative intensity :

587	23	627	36	724	15	804	71	868	50
928	36	955	47	1052	86	1089	34	1115	64
1156	96	1200	90	1260	96	1299	78	1354	55
1379	77	1446	95	1477	95	1498	100	1580	65
1611	24	1650	35	1958	13	2972	89	3023	36
3362	56								

COMPOUND : **Ethylene thiourea**

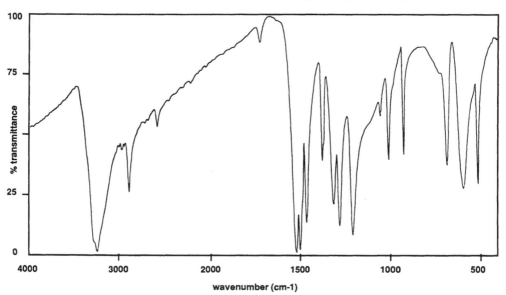

Chemical name   : ethylene thiourea

Other names   : ETU
Type   : fungicide
Brutoformula   : C3H6N2S   CAS nr   : 96-45-7
Molecular mass   : 102.15467   Exact mass   : 102.0251667
Instrument   : Bruker IFS-85   Optical resolution : 2 cm-1
Scans   : 32   Sampling technique : KBr pellet

Band maxima with relative intensity :

508	71	591	73	679	63	919	59	1001	61
1045	42	1205	92	1275	88	1308	79	1370	61
1461	87	1498	98	1520	100	1715	10	2570	45
2882	73	3245	98						

COMPOUND : **Etridiazole**

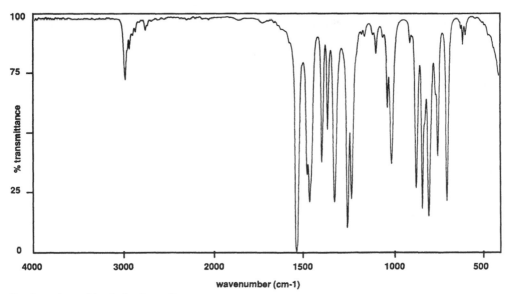

Chemical name : 5-ethoxy-3-(trichloromethyl)-1,2,4-thiadiazole

Other names : A Aterra, Terraclor, Terrazole, Ethazole, Olin 2424
Type : fungicide
Brutoformula : C5H5Cl3N2OS   CAS nr   : 2593-15-9
Molecular mass : 247.5274   Exact mass   : 245.9188155
Instrument : Bruker IFS-85   Optical resolution : 2 cm-1
Scans : 32   Sampling technique : neat film

Band maxima with relative intensity :

607	11	692	78	745	59	794	85	830	81
865	73	1008	62	1030	38	1092	15	1225	77
1246	89	1318	79	1358	48	1390	62	1455	79
1525	100	2984	27						

COMPOUND : **Etrimfos**

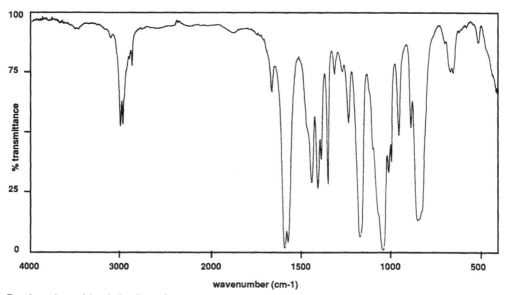

Chemical name	: O-6-ethoxy-2-ethylpirimidin-4-yl O,O-dimethyl phosphorothioate

Other names : Ekamet, SAN 197, Satisfar, 301074 E
Type : acaricide, insecticide
Brutoformula : C10H17N2O4PS   CAS nr   : 38260-54-7
Molecular mass : 292.29175   Exact mass   : 292.0646577
Instrument : Bruker IFS-85   Optical resolution : 2 cm-1
Scans : 32   Sampling technique : neat film

Band maxima with relative intensity :

505	12	645	25	846	87	881	48	947	51
988	62	1003	66	1035	100	1167	94	1227	45
1305	25	1345	71	1382	61	1402	73	1436	71
1589	98	1658	32	2849	20	2953	45	2980	46

COMPOUND : **Etrofolan**

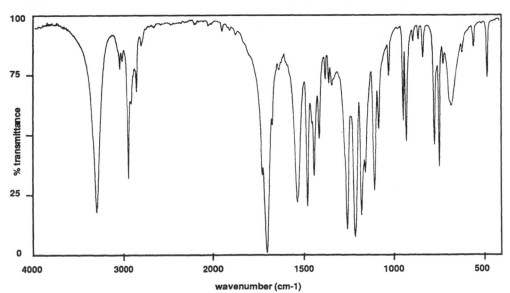

Chemical name : (2-isopropyl)-phenyl N-methylcarbamate

Other names : Bayer 105807
Type : insecticide
Brutoformula : C11H15NO2   CAS nr   : -
Molecular mass : 193.2477   Exact mass   : 193.1102704
Instrument : Bruker IFS-85   Optical resolution : 2 cm-1
Scans : 32   Sampling technique : KBr pellet

Band maxima with relative intensity :

485	26	559	13	683	38	750	64	776	54
837	17	862	9	934	53	949	44	1028	25
1084	47	1109	73	1161	66	1184	84	1219	93
1263	90	1346	29	1364	28	1383	26	1419	51
1449	67	1487	80	1542	78	1714	100	2868	31
2963	67	3059	21	3325	82				

COMPOUND : **Fenaminosulf**

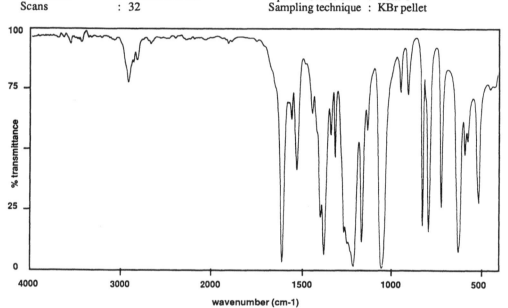

Chemical name : sodium -4-dimethylaminobenzenediazosulphonate

Other names : Lesan, Dexon, Bayer 22555
Type : fungicide
Brutoformula : C8H10N3O3SNa
Molecular mass : 251.23697
Instrument : Bruker IFS-85
Scans : 32

CAS nr : 140-56-7
Exact mass : 251.0340616
Optical resolution : 2 cm-1
Sampling technique : KBr pellet

Band maxima with relative intensity :

512	72	587	51	622	93	718	74	788	84
822	81	901	26	943	25	1050	100	1131	41
1165	88	1213	99	1315	53	1340	43	1379	94
1399	78	1444	34	1529	58	1558	37	1611	97
2918	21								

COMPOUND : **Fenamiphos**

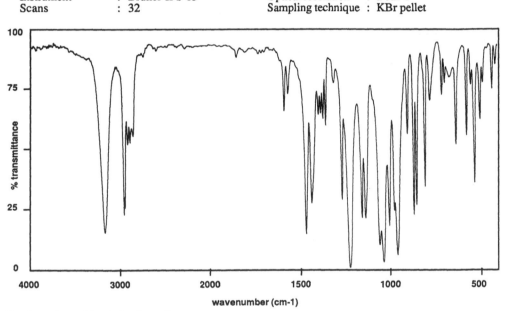

Chemical name	:	ethyl-4-methylthio-m-tolyl isopropylphosphoramidate		
Other names	:	Nemacur, Phenamiphos, Bayer 68138, SRA 3886		
Type	:	nematicide		
Brutoformula	:	C13H22NO3PS	CAS nr	: 22224-92-6
Molecular mass	:	303.35895	Exact mass	: 303.1057934
Instrument	:	Bruker IFS-85	Optical resolution	: 2 cm-1
Scans	:	32	Sampling technique	: KBr pellet

Band maxima with relative intensity :

421	14	439	24	491	21	506	37	539	64
557	22	581	44	640	47	702	21	718	26
784	29	811	65	857	73	872	77	909	43
965	94	1007	82	1039	97	1062	90	1141	79
1163	78	1231	100	1274	71	1320	21	1365	39
1380	36	1393	34	1406	35	1444	72	1476	85
1574	26	1596	33	2906	46	2934	47	2976	77
3189	84								

COMPOUND : **Fenarimol**

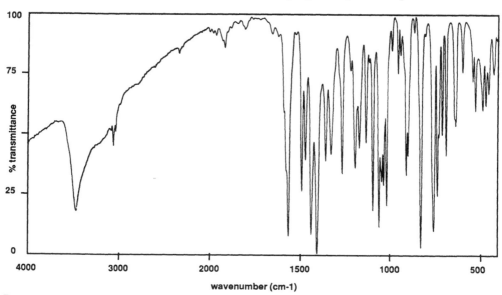

Chemical name	: 2,4'-dichloro-α-(pyrimidin-5-yl) benzhydryl alcohol

Other names	: EL 222, Rubigan, Bloc		
Type	: fungicide		
Brutoformula	: C17H12Cl2N2O	CAS nr	: 60168-88-9
Molecular mass	: 331.20399	Exact mass	: 330.0326619
Instrument	: Bruker IFS-85	Optical resolution	: 2 cm-1
Scans	: 32	Sampling technique	: KBr pellet

Band maxima with relative intensity :

423	23	451	32	467	37	485	39	524	40
539	27	595	23	633	46	685	58	706	50
732	75	755	90	826	97	860	6	896	58
906	66	937	16	950	23	982	14	1015	79
1032	71	1043	69	1058	88	1094	81	1131	53
1167	55	1191	63	1264	66	1326	57	1357	60
1406	100	1437	91	1469	60	1490	73	1564	92
3063	54	3465	81						

COMPOUND : **2-Chlorflurenol-methyl ester**

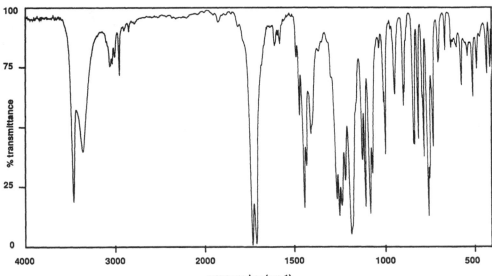

Chemical name : 2-chloro-9-hydroxyfluoren-9-carboxylic acid-methyl ester

Other names : -
Type :
Brutoformula : C15H11O3Cl   CAS nr   : 2536-31-4
Molecular mass : 274.7061   Exact mass   : 274.03966
Instrument : Bruker IFS-85   Optical resolution : 2 cm-1
Scans : 32   Sampling technique : KBr pellet

Band maxima with relative intensity :

426	28	481	26	505	38	568	33	624	17
657	18	696	23	729	59	758	88	781	63
811	55	838	57	894	41	943	36	999	62
1071	70	1086	87	1112	84	1128	64	1194	96
1223	73	1244	83	1258	88	1271	81	1409	53
1434	66	1449	84	1470	45	1577	14	1607	15
1720	99	1742	100	2951	27	3059	23	3369	59
3476	70								

189

COMPOUND : **Fenbutatin oxide**

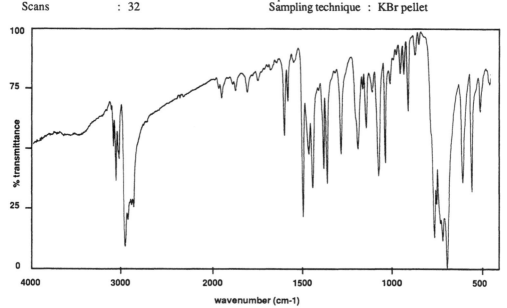

Chemical name	: bis [tris (2-methyl-2-phenylpropyl) tin] oxide		
Other names	: Torque, Vendex, SD 14114, Neostanox		
Type	: acaricide		
Brutoformula	: C60H78OSn2	CAS nr	: 13356-08-6
Molecular mass	: 1052.67006	Exact mass	: 1054.4096329
Instrument	: Bruker IFS-85	Optical resolution	: 2 cm-1
Scans	: 32	Sampling technique	: KBr pellet

Band maxima with relative intensity :

451	22	505	34	555	67	606	64	698	100
722	88	755	72	767	87	865	9	905	33
927	18	946	17	1002	21	1031	55	1072	60
1104	25	1137	40	1184	49	1278	51	1358	63
1377	57	1440	65	1461	51	1495	77	1579	28
1599	43	1808	24	1871	24	1947	27	2956	89
3021	52	3057	61	3085	46				

COMPOUND : **Fenchlorphos**

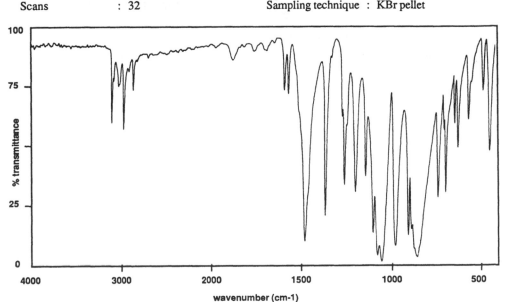

Chemical name : O,O-dimethyl O-(2,4,5-trichlorophenyl) phosphorothioate

Other names : Ronnel, Nankor, Korlan, Trolene, Ectoral
Type : insecticide
Brutoformula : C8H8Cl3O3PS   CAS nr   : 299-84-3
Molecular mass : 321.54392   Exact mass   : 319.8997351
Instrument : Bruker IFS-85   Optical resolution : 2 cm-1
Scans : 32   Sampling technique : KBr pellet

Band maxima with relative intensity :

402	14	435	52	473	27	552	39	609	51
628	41	676	70	720	72	834	97	887	88
960	93	1034	100	1084	87	1128	63	1184	70
1244	67	1350	80	1460	91	1558	29	1580	27
1871	14	2850	27	2954	44	3016	26	3089	41

COMPOUND : **Fenfuram**

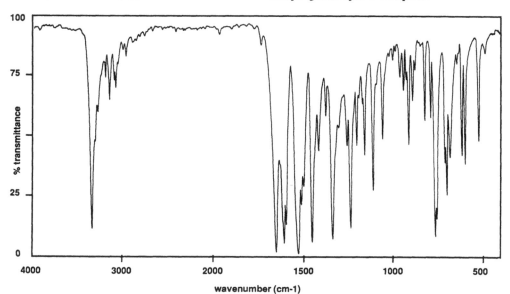

Chemical name : 2-methyl-3-furanilide

Other names : Panoram, GC401
Type : fungicide
Brutoformula : C12H11NO2   CAS nr   : 24691-80-3
Molecular mass : 201.22697   Exact mass   : 201.078972
Instrument : Bruker IFS-85   Optical resolution : 2 cm-1
Scans : 32   Sampling technique : KBr pellet

Band maxima with relative intensity :

508	53	583	62	600	59	667	59	686	75
695	62	753	93	774	43	808	44	874	35
897	54	924	31	944	25	1041	51	1098	73
1144	58	1189	54	1226	89	1244	54	1328	93
1362	41	1403	56	1442	94	1499	79	1518	100
1585	87	1597	95	1642	98	2921	15	3040	28
3111	33	3154	24	3317	88				

192

COMPOUND : **Fenitrothion**

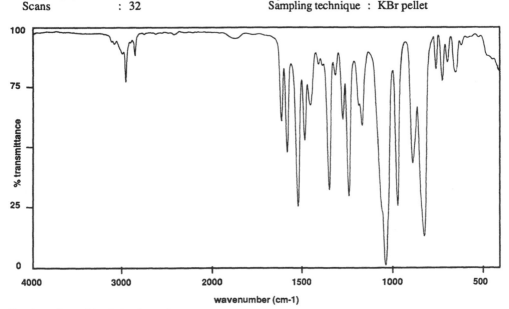

Chemical name   : O,O-dimethyl)-4-nitro-m-tolyl phosphorothioate

Other names   : Sulmithion, Folithion, Accothion, Novathion
Type   : acaricide, insecticide
Brutoformula   : C9H12NO5PS   CAS nr   : 122-14-5
Molecular mass   : 277.23345   Exact mass   : 277.0173756
Instrument   : Bruker IFS-85   Optical resolution : 2 cm-1
Scans   : 32   Sampling technique : KBr pellet

Band maxima with relative intensity :

648	18	692	13	721	21	756	16	829	87
889	56	972	74	1037	100	1163	40	1240	70
1271	37	1346	67	1450	31	1483	46	1523	74
1583	51	1616	38	2850	10	2954	21		

COMPOUND : **Fenoprop**

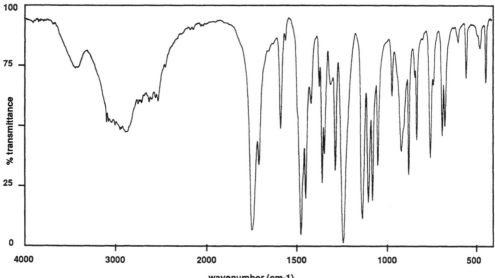

Chemical name   : 2-(2,4,5-trichlorophenoxy) propionic acid

Other names   : Kuran, Sylvex, 2,4,5-P
Type   : herbicide
Brutoformula   : C9H7Cl3O3   CAS nr   : 93-72-1
Molecular mass : 269.51334   Exact mass   : 267.9460738
Instrument   : Bruker IFS-85   Optical resolution : 2 cm-1
Scans   : 32   Sampling technique : KBr pellet

Band maxima with relative intensity :

441	32	476	17	553	30	597	15	670	50
685	54	752	63	828	56	873	71	915	61
967	37	1049	67	1079		1101	82	1134	89
1239	100	1283	69	1344	62	1356	74	1373	33
1417	40	1446	81	1472	96	1586	51	1705	66
1742	94	2884	52	3432	25				

COMPOUND : **Fenoxaprop-ethyl**

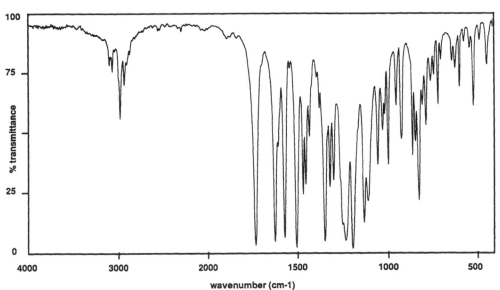

Chemical name   : ethyl-2[4(6-chloro-2-benzoxazolyl-oxy)]-phenoxypropanoate

Other names   : HOE 33171
Type   :
Brutoformula   : C18H16NO5Cl   CAS nr   : -
Molecular mass   : 361.7849   Exact mass   : 361.071690
Instrument   : Bruker IFS-85   Optical resolution : 2 cm-1
Scans   : 32   Sampling technique : KBr pellet

Band maxima with relative intensity :

402	11	440	21	481	11	516	39	595	31
620	23	638	20	700	19	716	38	739	25
757	28	784	47	803	38	824	79	843	54
861	60	923	53	950	39	995	64	1026	49
1054	64	1108	79	1131	88	1192	100	1230	96
1298	70	1319	73	1348	96	1378	40	1432	51
1452	72	1466	76	1505	99	1572	95	1628	96
1736	98	2940	30	2986	44	3073	24		

COMPOUND : **Fenoxycarb**

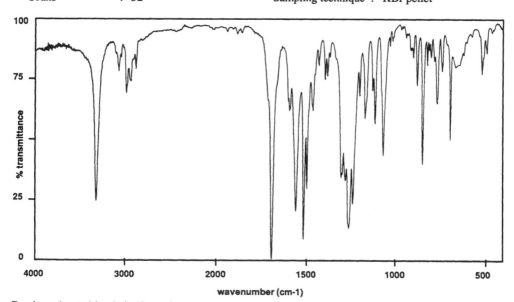

Chemical name : ethyl-[2-(4-phenoxyphenoxy) ethyl] carbamate

Other names : Insegar, Logic, Pictyl, RO 13-5223
Type : insecticide
Brutoformula : C17H19NO4   CAS nr   : 72490-01-8
Molecular mass : 301.34528   Exact mass   : 301.131397
Instrument : Bruker IFS-85   Optical resolution : 4 cm-1
Scans : 32   Sampling technique : KBr pellet

Band maxima with relative intensity :

516	21	690	49	734	20	763	34	798	14
817	17	842	59	873	26	894	14	1060	56
1107	42	1164	40	1195	31	1232	76	1253	86
1299	65	1377	22	1392	24	1425	18	1458	37
1490	70	1508	91	1552	79	1591	37	1687	100
2875	19	2979	30	3068	20	3313	75		

COMPOUND : **Fenpropimorph**

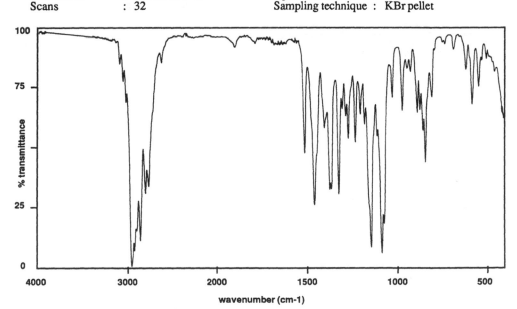

Chemical name : (±)-cis-4[3-(4-tert.butylphenyl)-2-methylpropyl]-2,6-dimethylmorpholine

Other names : Corbel, Mistral
Type : fungicide
Brutoformula : C20H33NO   CAS nr : 67306-03-0
Molecular mass : 303.49211   Exact mass : 303.2561991
Instrument : Bruker IFS-85   Optical resolution : 2 cm-1
Scans : 32   Sampling technique : KBr pellet

Band maxima with relative intensity :

405	35	536	24	573	31	608	17	675	8
797	28	835	56	865	34	880	35	917	17
965	34	1019	28	1084	94	1143	92	1178	40
1201	35	1230	47	1268	46	1322	69	1372	67
1400	41	1458	74	1511	52	2772	66	2807	69
2866	89	2965	100						

COMPOUND : **Fensulfothion**

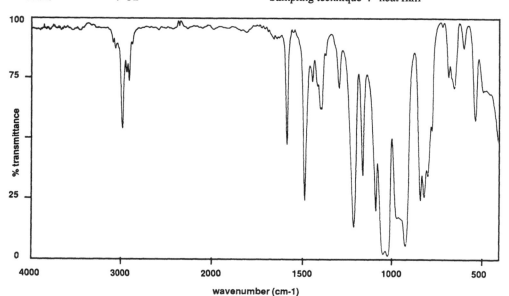

Chemical name : O,O-diethyl O-4-methylsulphinylphenyl phosphorothioate

Other names : Fensulfothion, Dasanit, Terracur, Bayer 25141
Type : nematicide, insecticide
Brutoformula : C11H17O4PS2   CAS nr   : 115-90-2
Molecular mass : 308.3495   Exact mass   : 308.030584
Instrument : Bruker IFS-85   Optical resolution : 2 cm-1
Scans : 32   Sampling technique : neat film

Band maxima with relative intensity :

404	55	532	43	593	12	649	29	820	75
841	76	925	95	1023	100	1089	80	1163	65
1214	87	1293	28	1392	37	1442	26	1489	76
1587	52	2907	24	2984	45				

COMPOUND : **Fenthion**

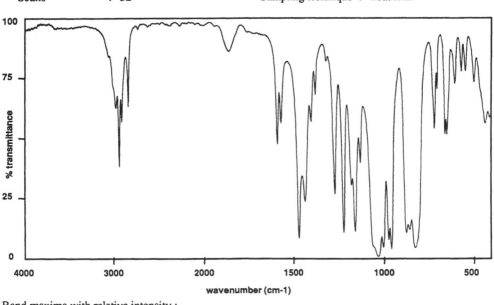

Chemical name : O,O-dimethyl O-4-methylthio-m-tolyl phosphorothioate

Other names : Lebacid, Baytex, Entex, Queletox, Tiguvon
Type : insecticide
Brutoformula : C10H15O3PS2   CAS nr   : 55-38-9
Molecular mass : 278.32301   Exact mass   : 278.0200207
Instrument : Bruker IFS-85   Optical resolution : 2 cm-1
Scans : 32   Sampling technique : neat film

Band maxima with relative intensity :

498	26	544	21	568	21	604	26	650	48
661	48	704	29	720	45	834	96	882	90
963	96	1035	100	1130	60	1161	89	1224	89
1273	73	1379	31	1403	42	1438	76	1474	92
1573	42	1594	52	1867	12	2846	35	2920	42
2949	61	2984	36						

COMPOUND : **Fentinchloride**

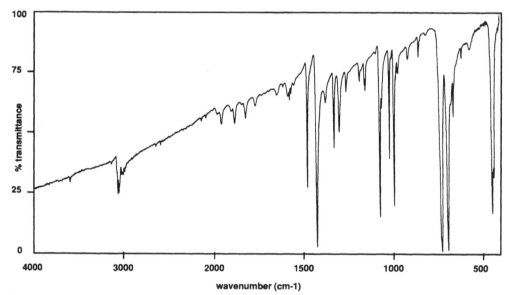

Chemical name   :   chloro-triphenyl stannane

Other names   :   Triphenylchlortin
Type   :   fungicide
Brutoformula   :   C18H15ClSn   CAS nr   :   639-58-7
Molecular mass   :   385.4632   Exact mass   :   385.988422
Instrument   :   Bruker IFS-85   Optical resolution   :   2 cm-1
Scans   :   32   Sampling technique   :   KBr pellet

Band maxima with relative intensity :

448	84	662	42	695	99	729	100	858	17
919	18	996	80	1021	60	1075	84	1157	31
1190	27	1262	31	1301	48	1331	55	1429	97
1480	72	1577	34	1824	42	1884	44	1959	44
3065	73								

COMPOUND : **Fentin (hydroxide)**

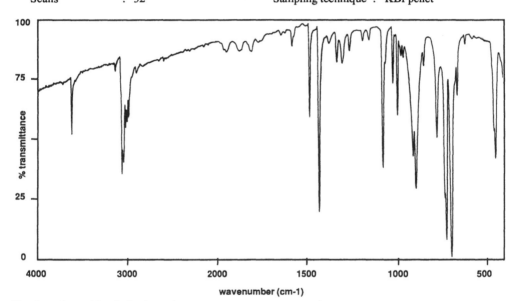

Chemical name   : triphenyltin hydroxide

Other names   : TPTH, Duter, Haitin, Tubotin
Type   : fungicide
Brutoformula   : C18H16OSn   CAS nr   : 76-87-9
Molecular mass   : 367.01762   Exact mass   : 368.0223077
Instrument   : Bruker IFS-85   Optical resolution : 2 cm-1
Scans   : 32   Sampling technique : KBr pellet

Band maxima with relative intensity :

447	58	661	31	694	100	723	92	775	49
848	18	895	70	911	56	997	39	1022	25
1077	61	1260	11	1300	17	1329	16	1428	80
1479	40	1578	9	2988	39	3022	44	3063	63
3616	47								

COMPOUND : **Fenvalerate**

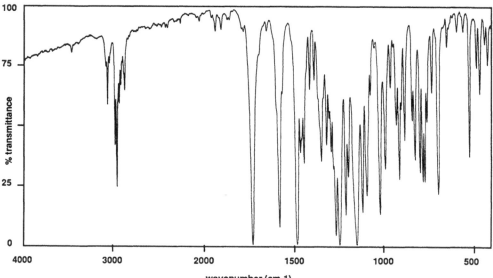

Chemical name : α-cyano-3-phenoxybenzyl 2-(4-chlorophenyl)-3-methylbutyrate

Other names : Sumicidine, Semi-alpha, Esfenvalerate, Pydrin
Type : acaricide, insecticide
Brutoformula : C25H22ClNO3   CAS nr   : 51630-58-1
Molecular mass : 419.91199   Exact mass   : 419.1288098
Instrument : Bruker IFS-85   Optical resolution   : 2 cm-1
Scans : 32   Sampling technique   : KBr pellet

Band maxima with relative intensity :

466	36	523	63	696	78	730	36	757	48
770	73	782	73	797	69	825	64	840	47
882	56	913	72	928	49	959	33	989	68
1018	87	1070	36	1091	79	1117	86	1150	100
1195	71	1211	87	1246	99	1268	95	1293	60
1319	57	1350	64	1389	30	1414	34	1446	65
1466	60	1488	99	1585	92	1737	99	2870	33
2912	31	2930	38	2957	74	2977	56	3061	39

202

COMPOUND : **Ferbam**

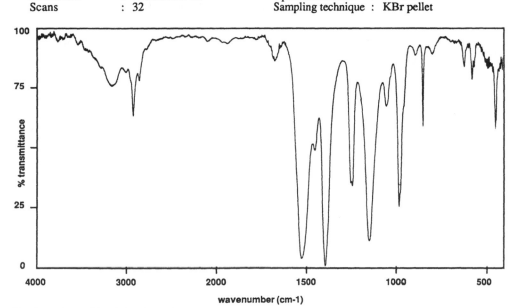

Chemical name   : iron-tris(dimethyldithiocarbamate)

Other names   : Trifungol, Ferberk, Hexaferb
Type   : fungicide
Brutoformula   : C9H18N3S6Fe   CAS nr   : 14484-64-1
Molecular mass   : 416.47091   Exact mass   : 415.9174366
Instrument   : Bruker IFS-85   Optical resolution : 2 cm-1
Scans   : 32   Sampling technique : KBr pellet

Band maxima with relative intensity :

444	41	573	20	618	15	848	40	977	74
1047	32	1145	89	1238	65	1391	100	1523	96
1678	12	2924	36	3165	23				

COMPOUND : **Fluazifop-butyl**

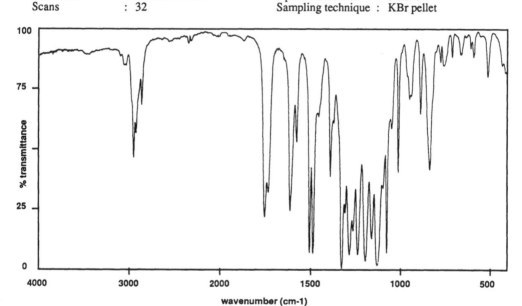

Chemical name   : butyl 2-[4-(5-trifluoromethyl-2-pyridyloxy) phenoxy] propionate

Other names	: Fusilade, Onecide, PP009		
Type	: herbicide		
Brutoformula	: C19H20F3NO4	CAS nr	: 69806-50-4
Molecular mass	: 383.37075	Exact mass	: 383.1344282
Instrument	: Bruker IFS-85	Optical resolution	: 2 cm-1
Scans	: 32	Sampling technique	: KBr pellet

Band maxima with relative intensity :

514	19	593	10	667	9	711	10	757	14
776	12	837	58	889	34	950	28	1011	59
1078	93	1131	98	1162	87	1196	96	1238	94
1284	94	1328	100	1393	61	1487	93	1506	93
1578	46	1615	75	1754	78	1958	15	2029	5
2875	30	2962	53						

COMPOUND : **Flubenzimine**

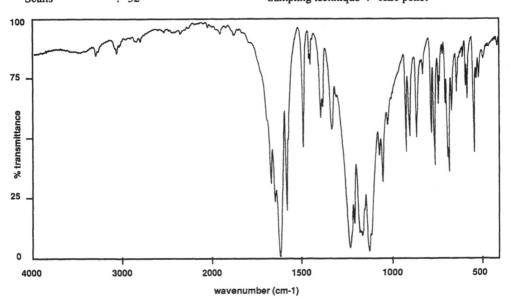

Chemical name	: N-[3-phenyl-4,5-bis[(trifluoromethyl) imino]-2-thiazolidinylidene] benzamine
Other names	: Fluthiamine, Cropotex, SLJ0312
Type	: acaricide, fungicide

Brutoformula	: C17H10F6N4S	CAS nr	: 37893-02-0
Molecular mass	: 416.34645	Exact mass	: 416.0530247
Instrument	: Bruker IFS-85	Optical resolution	: 2 cm-1
Scans	: 32	Sampling technique	: KBr pellet

Band maxima with relative intensity :

516	24	545	56	582	33	590	28	642	30
671	38	688	64	705	35	744	35	766	61
784	47	831	23	869	49	906	49	927	56
1057	68	1076	57	1137	98	1174	91	1216	86
1241	96	1334	46	1396	41	1451	18	1494	53
1591	80	1631	100	1679	68	3065	12		

COMPOUND : **Fluometuron**

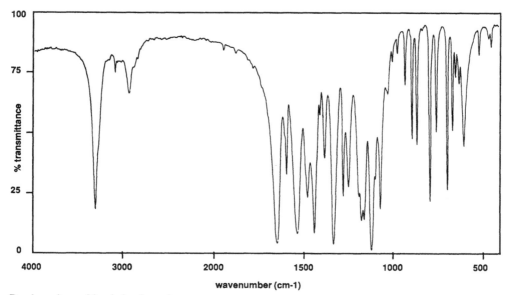

Chemical name : 1,1-dimethyl-3-(α,α,α-trifluoro-m-tolyl) urea

Other names	: Cotoran, C2059, Lanex		
Type	: herbicide		
Brutoformula	: C10H11F3N2O	CAS nr	: 2164-17-2
Molecular mass	: 232.20717	Exact mass	: 232.0823377
Instrument	: Bruker IFS-85	Optical resolution	: 2 cm-1
Scans	: 32	Sampling technique	: KBr pellet

Band maxima with relative intensity :

452	14	520	17	606	56	631	29	650	26
668	49	696	74	757	49	794	79	866	55
892	52	930	30	973	16	1070	82	1120	100
1176	87	1248	73	1279	77	1333	97	1384	60
1442	92	1482	77	1539	92	1599	67	1652	96
2934	33	3090	24	3317	82				

COMPOUND : **Fluorochloridone**

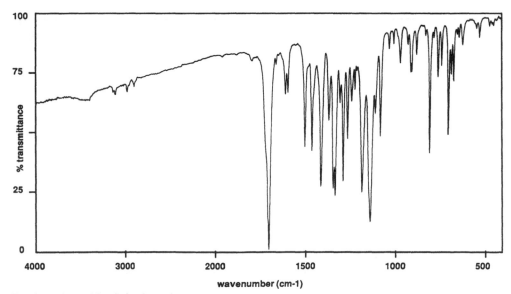

Chemical name   : 3-chloro-4-chloromethyl-1-(3-trifluoromethylphenyl)-2-pyrrolidinone

Other names   : Racer, R40244
Type   : herbicide
Brutoformula   : C12H10Cl2F3NO   CAS nr   : 61213-25-0
Molecular mass   : 312.1208   Exact mass   : 311.0091461
Instrument   : Bruker IFS-85   Optical resolution : 4 cm-1
Scans   : 32   Sampling technique : KBr pellet

Band maxima with relative intensity :

522	9	619	13	698	48	736	18	756	24
800	53	873	13	902	26	964	18	1026	8
1072	52	1128	95	1172	83	1232	38	1255	52
1280	69	1330	83	1361	46	1406	78	1456	57
1496	57	1608	27	1699	100	2893	7	2970	10
3101	10								

COMPOUND : **Flutolanil**

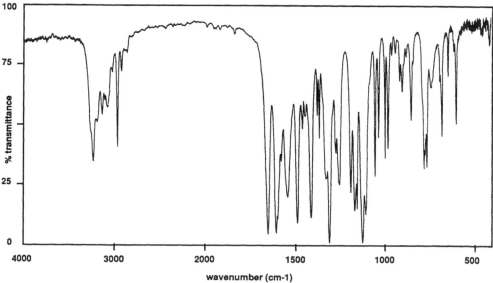

Chemical name : α,α,α-trifluoro-3-isopropoxy-o-toluanilide

Other names : -
Type :
Brutoformula : C17H16F3NO2   CAS nr   : -
Molecular mass : 323.31777   Exact mass   : 323.1133016
Instrument : Bruker IFS-85   Optical resolution : 2 cm-1
Scans : 32   Sampling technique : KBr pellet

Band maxima with relative intensity :

419	17	459	13	603	50	647	29	682	55
741	34	764	68	781	68	853	48	903	36
916	31	939	20	960	20	982	60	999	64
1035	55	1055	72	1110	88	1131	100	1160	85
1174	86	1196	78	1258	75	1279	62	1318	99
1373	55	1384	46	1419	89	1451	46	1464	51
1494	91	1547	80	1612	95	1658	96	2932	26
2979	58	3088	42	3149	45	3250	64		

COMPOUND : **Fluvalinate**

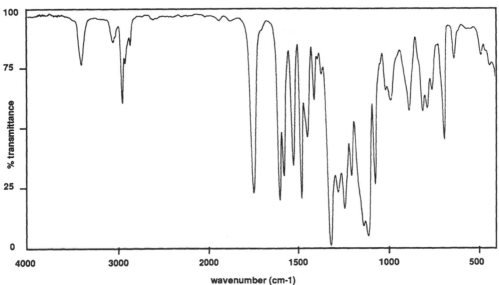

Chemical name : cyano (3-phenoxyphenyl) methyl N-[2-chloro-4-(trifluoromethyl) phenyl-D-valinate

Other names : Mavrik, Spur, Kartan, Aquaflow
Type : acaricide, insecticide
Brutoformula : C26H22ClF3N2O3
Molecular mass : 502.92504
Instrument : Bruker IFS-85
Scans : 32

CAS nr : 69409-94-5
Exact mass : 502.1270896
Optical resolution : 4 cm-1
Sampling technique : neat film

Band maxima with relative intensity :

634	21	694	55	784	41	811	43	891	43
991	38	1080	74	1120	96	1209	70	1247	84
1284	77	1326	100	1411	37	1452	53	1487	79
1533	65	1589	70	1612	80	1757	77	2968	38
3066	12	3408	21						

COMPOUND : **Folpet**

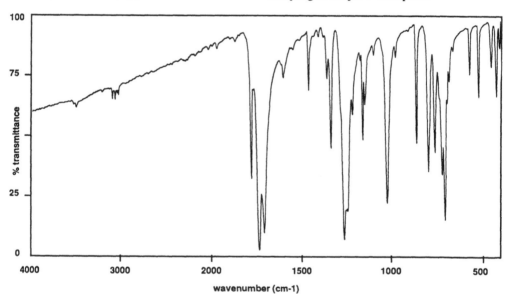

Chemical name   : N-(trichloromethylthio) phtalimide

Other names   : Phaltan, Folpan, Vinicoll
Type   : fungicide
Brutoformula   : C9H4Cl3NO2S   CAS nr   : 133-07-3
Molecular mass   : 296.55673   Exact mass   : 294.9028318
Instrument   : Bruker IFS-85   Optical resolution : 2 cm-1
Scans   : 32   Sampling technique : KBr pellet

Band maxima with relative intensity :

411	15	430	36	458	23	526	36	575	26
714	88	728	68	766	59	803	67	865	55
1027	80	1151	39	1163	53	1222	43	1271	95
1344	57	1365	27	1466	32	1609	27	1721	92
1748	100	1789	69						

COMPOUND : **Fonofos**

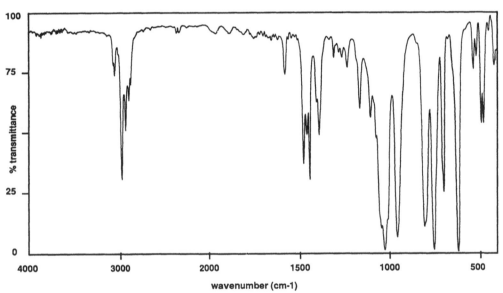

Chemical name   : O-ethyl S-phenyl ethylphosphonodithioate

Other names   : Dyfonate, N 2790
Type   : insecticide
Brutoformula   : C10H15OPS2   CAS nr   : 944-22-9
Molecular mass   : 246.32421   Exact mass   : 246.0301925
Instrument   : Bruker IFS-85   Optical resolution : 2 cm-1
Scans   : 32   Sampling technique : neat film

Band maxima with relative intensity :

402	24	411	21	472.	45	485	45	510	17
527	22	614	100	690	74	746	99	797	89
950	93	1020	99	1098	42	1158	39	1227	21
1303	17	1387	50	1439	69	1455	49	1473	62
1577	24	2897	35	2935	47	2978	68	3058	24

COMPOUND : **Formetanate**

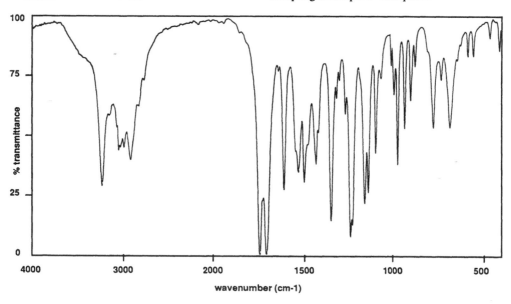

Chemical name   : 3-dimethylamino-methyleneiminophenyl methylcarbamate

Other names	: -		
Type	: acaricide, insecticide		
Brutoformula	: C11H15N3O2	CAS nr	: 22259-30-9
Molecular mass	: 221.2611	Exact mass	: 221.1164168
Instrument	: Bruker IFS-85	Optical resolution	: 2 cm-1
Scans	: 32	Sampling technique	: KBr pellet

Band maxima with relative intensity :

409	14	460	8	550	16	580	16	678	46
725	26	769	46	867	20	892	35	925	47
963	62	983	32	1001	20	1092	57	1136	73
1156	78	1238	92	1268	40	1348	85	1431	61
1496	69	1528	65	1609	72	1705	99	1740	100
2915	59	3049	55	3233	70				

COMPOUND : **Formothion**

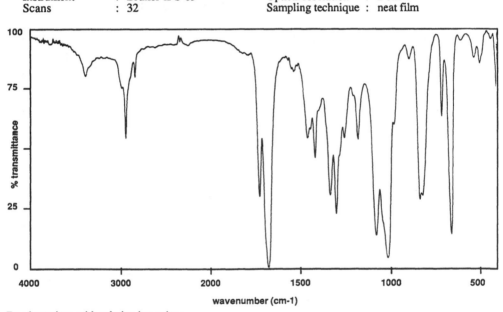

| Chemical name | : S-(N-formyl-N-methylcarbamoylmethyl) O,O-dimethyl phosphorodithioate |
| | |

Other names : Anthio, Aflix
Type : acaricide, insecticide
Brutoformula : C6H12NO4PS2   CAS nr   : 2540-82-1
Molecular mass : 257.2606   Exact mass   : 256.9945342
Instrument : Bruker IFS-85   Optical resolution : 2 cm-1
Scans : 32   Sampling technique : neat film

Band maxima with relative intensity :

402	34	496	13	657	86	710	36	835	71
893	11	1013	96	1078	86	1177	45	1298	77
1331	69	1416	53	1457	45	1676	100	1726	69
2844	18	2949	44						

COMPOUND : **Fuberdiazole**

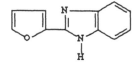

Chemical name   : 2-(2'-furyl) benzimidazole

Other names   : Furidazol, Voronit, Bayer 33172
Type   : fungicide
Brutoformula   : C11H8N2O   CAS nr   : 3878-19-1
Molecular mass   : 184.19921   Exact mass   : 184.0636573
Instrument   : Bruker IFS-85   Optical resolution : 2 cm-1
Scans   : 32   Sampling technique : KBr pellet

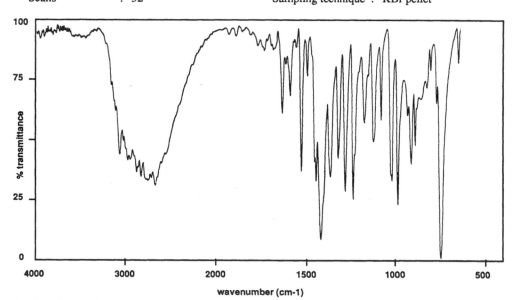

Band maxima with relative intensity :

641	17	738	100	762	34	795	20	883	52
905	59	979	77	1011	67	1075	41	1119	50
1169	42	1234	74	1278	71	1319	57	1364	65
1416	91	1442	66	1490	22	1525	62	1586	30
1629	38	2664	68	2823	64	3059	55		

COMPOUND : **Furalaxyl**

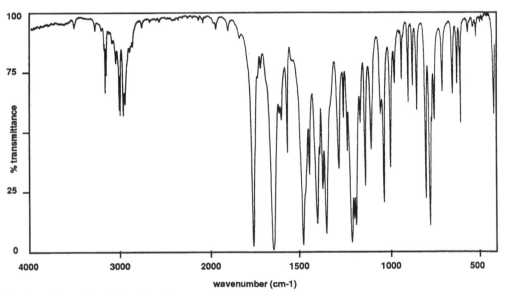

Chemical name   : methyl N-(2-furoyl)-N-(2,6-xylyl)-DL-alaninate

Other names   : CGA 38140, Fonganil, Fongarid
Type   : fungicide
Brutoformula   : C17H19NO4   CAS nr   : 57646-30-7
Molecular mass   : 301.34528   Exact mass   : 301.131397
Instrument   : Bruker IFS-85   Optical resolution : 2 cm-1
Scans   : 32   Sampling technique : KBr pellet

Band maxima with relative intensity :

593	46	611	30	638	34	696	33	741	45
768	90	790	78	838	41	862	30	887	37
925	27	966	29	991	65	1026	80	1043	42
1094	57	1126	73	1153	46	1178	89	1189	87
1202	97	1225	58	1245	44	1272	65	1343	93
1362	74	1392	89	1433	68	1469	97	1556	58
1589	44	1636	100	1705	22	1751	98	2949	42
2984	42	2997	39	3031	20	3134	25	3145	32

COMPOUND : **Fenpropathrin**

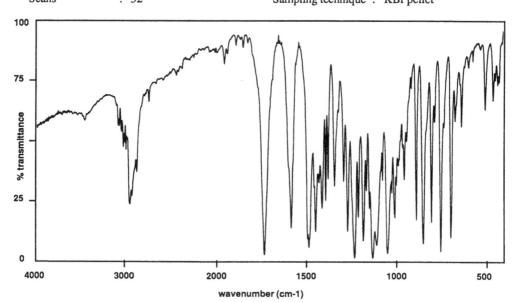

| Chemical name | : (RS)-α-cyano-3-phenoxybenzyl-2,2,3,3-tetramethylcyclopropane- |
| | carboxylate |

Other names : Fenpropanate, WL 41706
Type : insecticide, acaricide

Brutoformula	: C22H23NO3	CAS nr	: 64257-84-7
Molecular mass	: 349.4335	Exact mass	: 349.1677813
Instrument	: Bruker IFS-85	Optical resolution	: 2 cm-1
Scans	: 32	Sampling technique	: KBr pellet

Band maxima with relative intensity :

433	27	459	33	503	37	593	19	633	44
669	42	695	91	751	97	782	42	804	84
849	93	886	83	913	28	939	50	952	66
983	60	1004	82	1022	72	1046	97	1076	67
1108	94	1130	100	1148	79	1164	71	1181	92
1209	82	1229	99	1268	88	1291	67	1342	69
1378	66	1391	75	1411	78	1428	68	1447	88
1484	95	1583	87	1733	98	2868	63	2941	77
2984	54	3001	51	3014	53	3036	46	3060	44

COMPOUND : **Gibberellic acid**

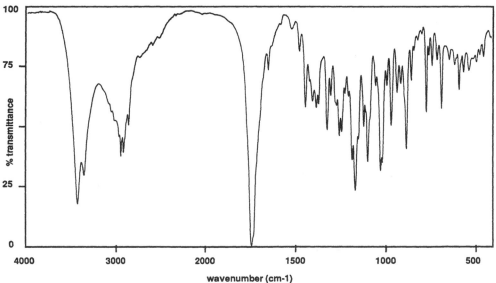

Chemical name	:	(3S,3aS,4S,4aS,7S,9aR,9bR,12S)-7,12-dihydroxy-3-methyl-6-methylene-2-oxoperhydro-4a,7-methano-9b,3-propeno [1,2-b]furan-4-carboxylic acid
Other names	:	Gibberillin, Berelex, Activol
Type	:	growth regulator
Brutoformula	:	C19H24O6
Molecular mass	:	348.3995
Instrument	:	Bruker IFS-85
Scans	:	32

CAS nr	:	77-06-5
Exact mass	:	348.15727
Optical resolution	:	2 cm-1
Sampling technique	:	KBr pellet

Band maxima with relative intensity :

459	20	541	26	569	26	595	34	692	42
717	21	745	23	778	43	863	30	892	59
922	31	943	35	975	49	998	32	1032	68
1104	64	1127	49	1173	76	1191	63	1250	51
1263	53	1310	36	1330	50	1391	41	1452	41
1484	17	1657	25	1752	100	2938	60	2968	61
3376	69	3448	81						

COMPOUND : **Glufosinate-ammonium**

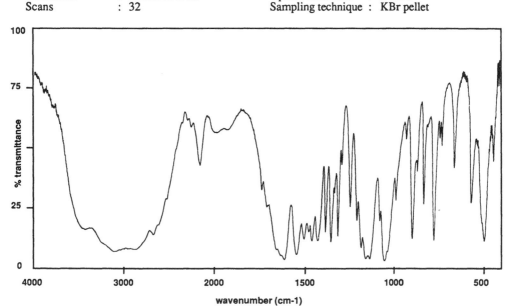

Chemical name   : ammonium (DL-homoalanine-4-yl) methylphosphinate

Other names	: HOE 39866, Basta		
Type	: herbicide		
Brutoformula	: C5H15N2O4P	CAS nr	: 77182-82-2
Molecular mass	: 198.16006	Exact mass	: 198.0769358
Instrument	: Bruker IFS-85	Optical resolution	: 2 cm-1
Scans	: 32	Sampling technique	: KBr pellet

Band maxima with relative intensity :

415	25	434	57	480	9	553	75	647	60
715	50	728	47	757	91	815	75	880	90
918	48	972	74	1037	100	1120	99	1193	82
1232	77	1281	59	1300	89	1341	92	1372	88
1415	91	1446	91	1491	91	1531	97	1598	99
1975	46	2139	60	3083	97				

COMPOUND : **Glyphosate**

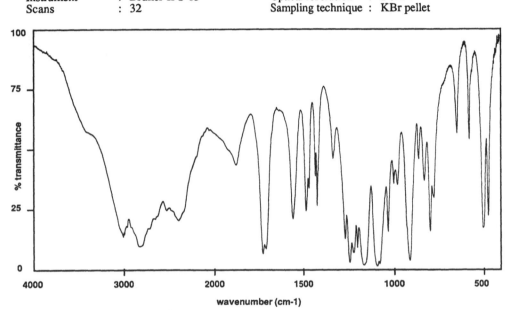

Chemical name   : N-(phosphonomethyl) glycine

Other names   : Roundup, CP 67573, MON 0573
Type   : herbicide
Brutoformula   : C3H8NO5P   CAS nr   : 1071-83-6
Molecular mass   : 169.07467   Exact mass   : 169.0140045
Instrument   : Bruker IFS-85   Optical resolution : 2 cm-1
Scans   : 32   Sampling technique : KBr pellet

Band maxima with relative intensity :

472	79	502	84	579	46	648	44	798	85
830	64	863	54	915	97	980	65	1001	65
1030	85	1091	100	1165	99	1202	92	1245	98
1270	88	1336	54	1422	74	1433	61	1485	76
1561	79	1732	94	1884	56	2408	80	2837	91
3012	86								

COMPOUND : **HCH (alpha)**

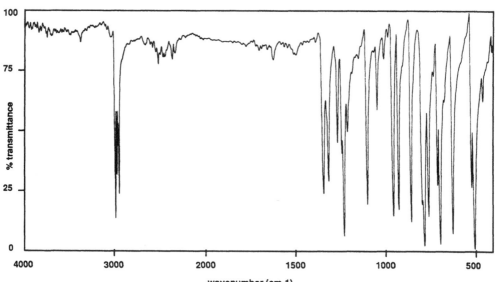

Chemical name : 1,2,3,4,5,6-hexachloro-cyclohexane (alpha isomer)

Other names	: -		
Type	: insecticide		
Brutoformula	: C6H6Cl6	CAS nr	: 319-84-6
Molecular mass	: 290.83272	Exact mass	: 287.8600662
Instrument	: Bruker IFS-85	Optical resolution	: 2 cm-1
Scans	: 32	Sampling technique	: KBr pellet

Band maxima with relative intensity :

456	37	507	100	520	73	624	93	694	97
706	72	758	86	784	98	856	88	924	83
954	85	1003	19	1043	40	1101	80	1208	49
1229	94	1240	58	1264	54	1315	70	1344	76
1492	17	1620	19	2362	18	2515	20	2948	75
2973	67	2991	85	3361	11				

COMPOUND : **HCH (beta)**

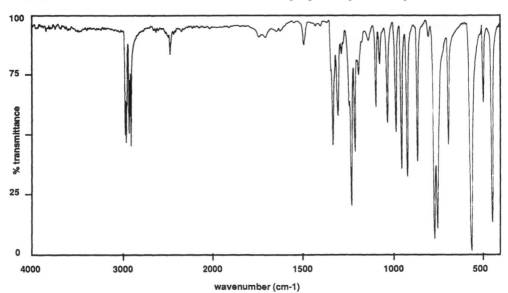

Chemical name	: 1,2,3,4,5,6-hexachlorocyclohexane (beta isomer)		
Other names	: -		
Type	: insecticide		
Brutoformula	: C6H6Cl6	CAS nr	: 319-85-7
Molecular mass	: 290.83272	Exact mass	: 287.8600662
Instrument	: Bruker IFS-85	Optical resolution	: 2 cm-1
Scans	: 32	Sampling technique	: KBr pellet

Band maxima with relative intensity :

448	88	495	37	564	100	689	54	753	90
770	94	864	61	922	68	955	65	985	49
1032	45	1075	20	1097	38	1192	24	1216	57
1236	80	1307	42	1334	54	1491	12	1707	8
2920	54	2932	41	2944	48	2976	52	2985	49

COMPOUND : **HCH (delta)**

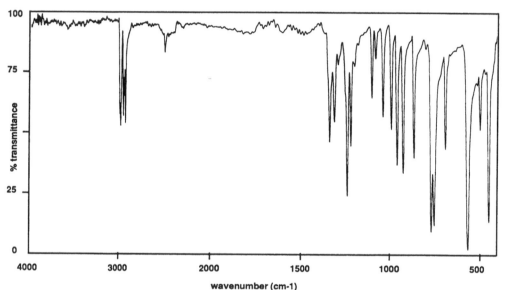

Chemical name	: 1,2,3,4,5,6-hexachlorocyclohexane (delta isomer)

Other names	: -		
Type	: insecticide		
Brutoformula	: C6H6Cl6	CAS nr	: 319-86-8
Molecular mass	: 290.83272	Exact mass	: 287.8600662
Instrument	: Bruker IFS-85	Optical resolution	: 2 cm-1
Scans	: 32	Sampling technique	: KBr pellet

Band maxima with relative intensity :

448	88	495	49	564	100	689	57	753	89
770	92	864	61	922	67	955	64	985	49
1032	43	1075	19	1097	35	1216	55	1236	77
1307	45	1334	54	2920	46	2932	34	2944	43
2975	47								

COMPOUND : **alpha-HEPO**

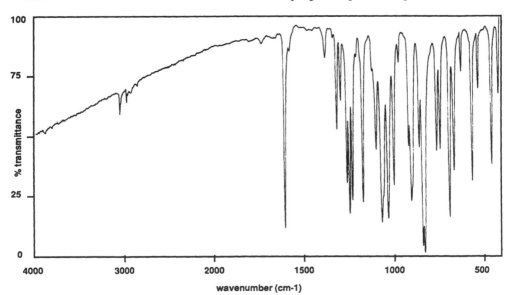

Chemical name   : 1,4,5,6,7,8,8-heptachloro-2,3-epoxy-3a,4,7,7a-tetrahydro 4,7-
                       methanoindene

Other names	: -		
Type	: metabolite of Heptachlor		
Brutoformula	: C10H5Cl7O	CAS nr	: 76-45-8
Molecular mass	: 389.3217	Exact mass	: 385.816008
Instrument	: Bruker IFS-85	Optical resolution	: 2 cm-1
Scans	: 32	Sampling technique	: KBr pellet

Band maxima with relative intensity :

413	33	453	62	528	30	563	69	621	23
662	65	687	85	737	56	757	57	827	100
838	97	856	55	902	78	917	54	969	19
999	71	1031	85	1066	87	1097	56	1172	78
1231	77	1246	82	1260	70	1293	35	1315	47
1379	16	1607	88	2976	34	3051	39		

COMPOUND : **Heptachlor**

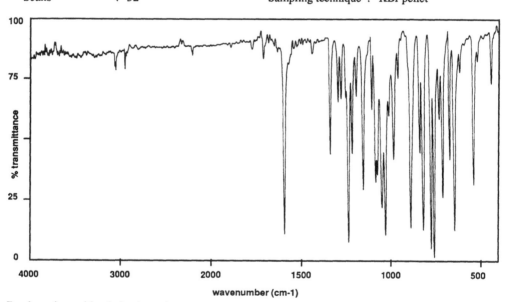

Chemical name	:	1,4,5,6,7,8,8-heptachloro--3a,4,7,7a-tetrahydro 4,7-methanoindene		

Other names	: Velsicol, Drinox, Heptagran, Heptox			
Type	: insecticide			
Brutoformula	: C10H5Cl7	CAS nr	: 76-44-8
Molecular mass	: 373.32235	Exact mass	: 369.8210947
Instrument	: Bruker IFS-85	Optical resolution	: 2 cm-1
Scans	: 32	Sampling technique	: KBr pellet

Band maxima with relative intensity :

450	26	549	69	622	21	655	88	681	58
720	74	738	41	770	100	787	96	831	88
848	55	899	87	967	24	992	58	1020	39
1041	90	1061	78	1087	64	1097	67	1116	37
1167	71	1205	31	1228	55	1251	93	1289	32
1305	33	1351	55	1450	13	1605	89	1719	15
2971	19	3075	19						

COMPOUND : **Heptenophos**

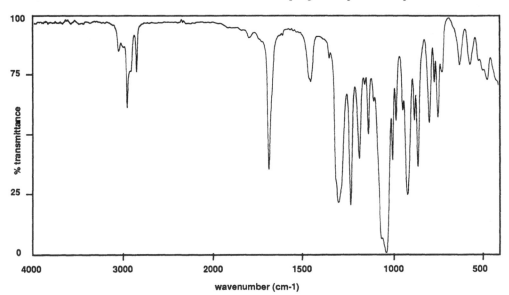

Chemical name : 7-chlorobicyclo[3.2.0]hepta-2,6-dien-6-yl dimethyl phosphate

Other names : Hostaquick, Ragadan, HOE 2982,W 13787
Type : insecticide, acaricide
Brutoformula : C9H12ClO4P   CAS nr : 23560-59-0
Molecular mass : 250.6203   Exact mass : 250.016168
Instrument : Bruker IFS-85   Optical resolution : 2 cm-1
Scans : 32   Sampling technique : KBr pellet

Band maxima with relative intensity :

468	26	563	19	622	20	743	42	763	27
791	44	856	63	876	43	918	75	981	43
1001	60	1036	100	1136	49	1185	59	1231	79
1297	78	1451	27	1683	64	2855	22	2958	37

225

COMPOUND : **HCB**

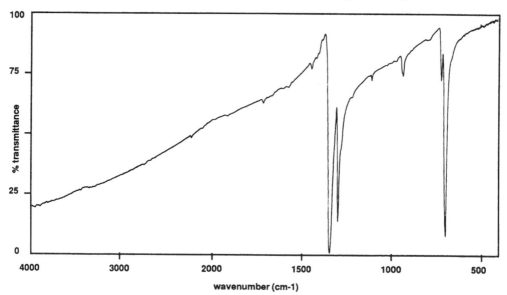

Chemical name   : hexachlorobenzene

Other names   : Anticarie, Bent-cure, Ceku-B, perchlorobenzene
Type   : fungicide
Brutoformula   : C6Cl6   CAS nr   : 118-74-1
Molecular mass   : 284.7849   Exact mass   : 281.8131186
Instrument   : Bruker IFS-85   Optical resolution : 2 cm-1
Scans   : 32   Sampling technique : KBr pellet

Band maxima with relative intensity :

696   92   721   26   932   24   1297   86   1342   100

COMPOUND : **Hexaconazole**

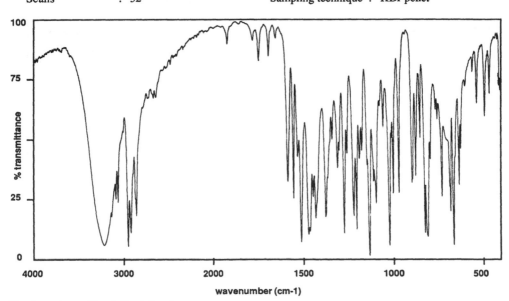

Chemical name   : 1-(N-1,2,4-triazol)-2-(2,4-dichlorophenyl)-hexanol-2

Other names   : PP 523, R 15423
Type   : fungicide
Brutoformula   : C14H17N3OCl2   CAS nr   :
Molecular mass   : 314.21709   Exact mass   : 313.0748581
Instrument   : Bruker IFS-85   Optical resolution : 4 cm-1
Scans   : 32   Sampling technique : KBr pellet

Band maxima with relative intensity :

620	55	629	70	661	95	676	81	726	75
749	42	789	59	807	92	822	90	848	50
875	66	896	69	968	73	999	73	1020	95
1053	45	1094	77	1134	100	1174	55	1188	58
1205	88	1220	83	1257	56	1273	90	1309	62
1340	50	1377	83	1433	83	1445	74	1471	90
1514	93	1531	57	1555	75	1586	68	2861	81
2927	88	2958	94	3067	76	3091	74	3225	94

COMPOUND : **Hexazinone**

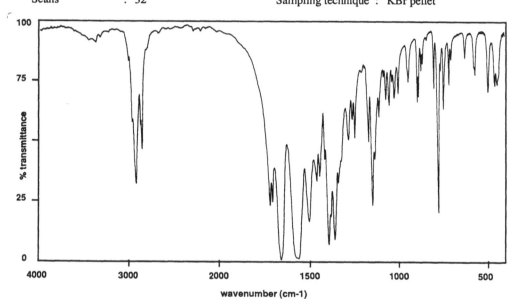

Chemical name	:	3-cyclohexyl-6-dimethylamino-1-methyl-1,3,5-triazine-2,4-dione

Other names : Velpar, DPX 3674
Type : herbicide
Brutoformula : C12H20N4O2   CAS nr : 51235-04-2
Molecular mass : 252.3188   Exact mass : 252.158613
Instrument : Bruker IFS-85   Optical resolution : 2 cm-1
Scans : 32   Sampling technique : KBr pellet

Band maxima with relative intensity :

461	27	499	29	570	22	626	15	717	26
750	36	783	80	804	27	875	20	891	31
897	33	947	25	1003	29	1027	31	1056	34
1077	32	1117	39	1157	77	1175	50	1253	48
1267	40	1290	49	1371	91	1404	93	1451	64
1467	66	1511	83	1569	99	1666	100	1711	75
1725	76	2859	52	2928	66				

COMPOUND : **Hexythiazox**

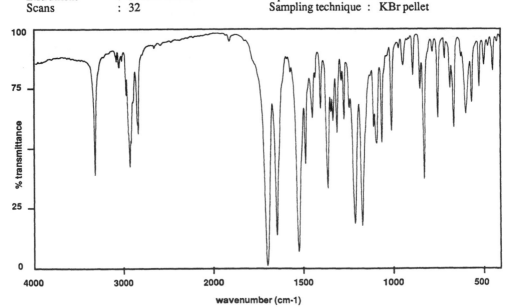

Chemical name : trans-5-(4-chlorophenyl-N-cyclohexyl-4-methyl-2-oxothiazolidine-3-carboxamide

Other names	: NA 73		
Type	: herbicide		
Brutoformula	: C17H15ClN2O2S	CAS nr	: -
Molecular mass	: 346.8343	Exact mass	: 346.0542694
Instrument	: Bruker IFS-85	Optical resolution	: 2 cm-1
Scans	: 32	Sampling technique	: KBr pellet

Band maxima with relative intensity :

447	18	497	15	524	25	566	31	599	36
666	42	686	25	716	12	755	38	783	10
830	64	850	25	891	19	946	15	1010	43
1062	48	1091	48	1108	41	1174	83	1215	82
1272	38	1313	44	1335	39	1347	35	1365	67
1405	33	1451	37	1490	57	1530	94	1653	87
1708	100	2844	43	2938	57	2979	27	3057	15
3333	60								

COMPOUND : **HOE 070542**

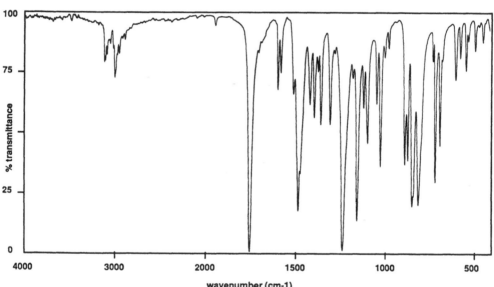

Chemical name : ethyl-1-(2,4-dichlorophenyl)-5-trifluoromethyl-1H-1,2,4-triazol-3-yl-carboxylate

Other names : -
Type : herbicide
Brutoformula : C12H8Cl5N3O2   CAS nr : -
Molecular mass : 403.48146   Exact mass : 400.9059101
Instrument : Bruker IFS-85   Optical resolution : 2 cm-1
Scans : 32   Sampling technique : KBr pellet

Band maxima with relative intensity :

443	10	486	14	537	22	568	17	595	26
679	54	705	70	719	18	801	79	836	80
860	60	877	62	968	13	1015	63	1037	37
1089	53	1111	38	1146	86	1225	99	1297	45
1349	45	1364	23	1384	42	1408	37	1472	82
1500	33	1570	24	1587	31	1738	100	2986	26
3103	20								

COMPOUND : **Hydroxy-atrazine**

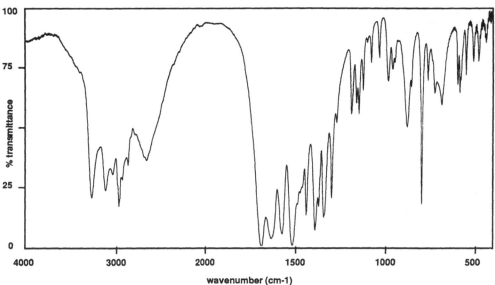

Chemical name	:	4-(ethylamino)-6-[(isopropyl)amino]-1,3,5-triazin-2(1H)-one		
Other names	:	-		
Type	:	metabolite of Atrazine		
Brutoformula	:	C8H15N5O	CAS nr	: 2163-68-0
Molecular mass	:	197.2416	Exact mass	: 197.127649
Instrument	:	Bruker IFS-85	Optical resolution	: 2 cm-1
Scans	:	32	Sampling technique	: KBr pellet

Band maxima with relative intensity :

439	13	479	21	507	21	546	27	582	34
594	31	682	40	759	29	792	82	877	49
958	24	981	29	1030	20	1077	22	1122	33
1145	43	1159	39	1187	44	1297	79	1341	87
1388	93	1436	86	1515	99	1569	94	1630	96
1685	100	2659	63	2971	83	3119	76	3269	79

COMPOUND : **Imazalil**

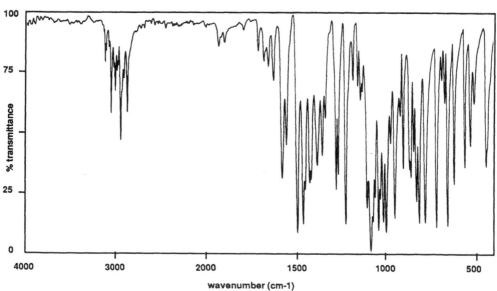

Chemical name   : 1-(β-allyloxy-2,4-dichlorophenylethyl) imidazole

Other names   : Fungaflor, R 23979, Bromazil
Type   : fungicide
Brutoformula   : C14H14Cl2N2O   CAS nr   : 35554-44-0
Molecular mass   : 297.18648   Exact mass   : 296.0483111
Instrument   : Bruker IFS-85   Optical resolution : 2 cm-1
Scans   : 32   Sampling technique : KBr pellet

Band maxima with relative intensity :

450	64	514	37	535	55	566	64	623	71
659	89	675	37	722	89	785	88	817	88
832	79	846	57	862	68	903	65	919	42
949	86	972	54	996	91	1011	87	1039	91
1060	77	1085	100	1106	81	1143	36	1159	30
1229	88	1271	67	1283	73	1345	43	1362	59
1390	63	1433	71	1468	88	1501	92	1563	54
1587	69	2874	40	2947	52	3008	31	3058	41

COMPOUND : **Imazamethabenz-methyl(mix)**

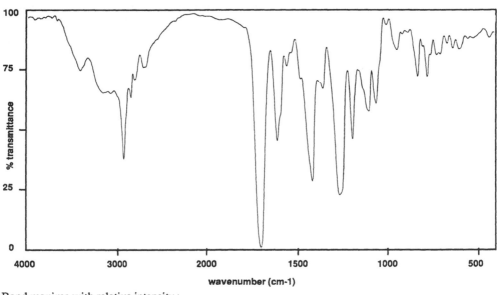

Chemical name	:	2-(4-isopropyl-4-methyl-5-oxo-2-imidazolin-2-yl)-methyl-benzoic acid-methyl ester (mixture of meta and para methyl)		
Other names	:	AC-222.293, Assert, Dagger		
Type	:	herbicide		
Brutoformula	:	C16H20N2O3	CAS nr	: 81405-85-8
Molecular Mass	:	288.3494	Exact mass	: 288.147380
Instrument	:	Bruker IFS-85	Optical resolution	: 4 cm-1
Scans	:	32	Sampling technique	: KBr pellet

Band maxima with relative intensity :

530	13	636	20	732	23	791	32	833	33
947	17	1064	42	1107	41	1209	76	1300	80
1371	36	1436	76	1564	30	1629	60	1728	100
2715	24	2970	65						

233

COMPOUND : **Imazamethabenz-methyl (para)**

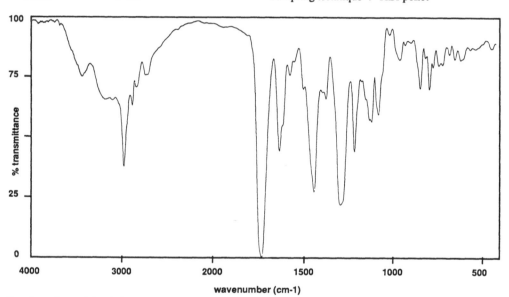

Chemical name	:	2-(4-isopropyl-4-methyl-5-oxo-2-imidazolin-2-yl)-4-methyl-benzoic acid methyl ester		
Other names	:	AC-4094-8-(5)		
Type	:	herbicide		
Brutoformula	:	C16H20N2O3	CAS nr	: -
Molecular Mass	:	288.3494	Exact mass	: 288.147380
Instrument	:	Bruker IFS-85	Optical resolution	: 4 cm-1
Scans	:	32	Sampling technique	: KBr pellet

Band maxima with relative intensity :

687	30	701	32	772	73	793	65	830	60
1023	100	1183	45	1257	87	1457	43	1478	36
1583	13	2849	21	2952	44				

COMPOUND : **Imazapyr**

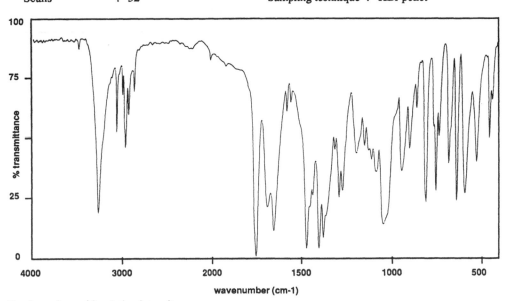

Chemical name : 2-(4-isopropyl-4-methyl-5-oxo-2-imidazolin-2-yl) nicotinic acid
Other names   : AC-243.997 , Arsenal, Cl-252.925
Type   : herbicide
Brutoformula   : C13H15N3O3   CAS nr   : 81334-34-1
Molecular Mass : 244.27543   Exact mass   : 244.1085922
Instrument   : Bruker IFS-85   Optical resolution : 2 cm-1
Scans   : 32   Sampling technique : KBr pellet

Band maxima with relative intensity :

452	49	522	59	586	72	632	76	677	60
730	49	747	72	803	76	857	37	897	54
940	63	1040	86	1080	63	1146	53	1191	56
1266	72	1286	75	1311	55	1373	92	1397	96
1465	96	1560	35	1582	39	1652	89	1689	79
1752	100	2874	30	2936	40	2968	54	3003	32
3066	48	3262	82						

COMPOUND : **Iodofenphos**

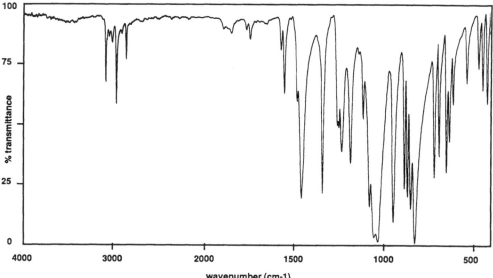

Chemical name   : O-2,5-dichloro-4-iodophenyl O,O-dimethyl phosphorothioate

Other names   : Nuvanol, Jodfenphos
Type   : insecticide, acaricide
Brutoformula   : C8H8Cl2IO3PS   CAS nr   : 18181-70-9
Molecular mass   : 412.99542   Exact mass   : 411.835542
Instrument   : Bruker IFS-85   Optical resolution : 2 cm-1
Scans   : 32   Sampling technique : KBr pellet

Band maxima with relative intensity :

419	41	444	35	467	26	532	32	608	41
630	57	647	70	686	63	713	72	825	100
846	85	864	80	881	76	943	91	1028	99
1076	84	1109	47	1180	65	1228	61	1244	50
1335	78	1452	80	1549	36	1568	18	1741	13
1845	10	2847	21	2954	40	2999	14	3076	31

COMPOUND : **Ioxynil**

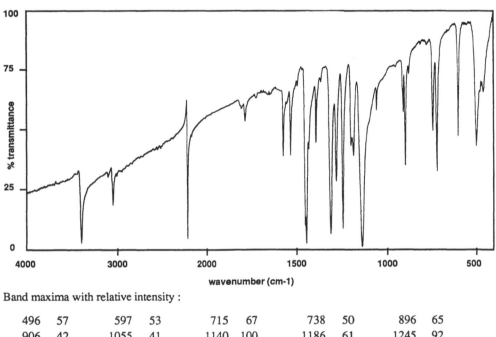

Chemical name	: 4-hydroxy-3,5-diiodobenzonitrile		
Other names	: Oxytril, Certrol, Actril, Bentrol		
Type	: herbicide		
Brutoformula	: C7H3I2NO	CAS nr	: 1689-83-4
Molecular mass	: 370.91706	Exact mass	: 370.8307811
Instrument	: Bruker IFS-85	Optical resolution	: 2 cm-1
Scans	: 32	Sampling technique	: KBr pellet

Band maxima with relative intensity :

496	57	597	53	715	67	738	50	896	65
906	42	1055	41	1140	100	1186	61	1245	92
1280	71	1312	94	1392	55	1446	98	1533	60
1574	61	1786	46	2225	96	3057	81	3401	97

COMPOUND : **Iprodione**

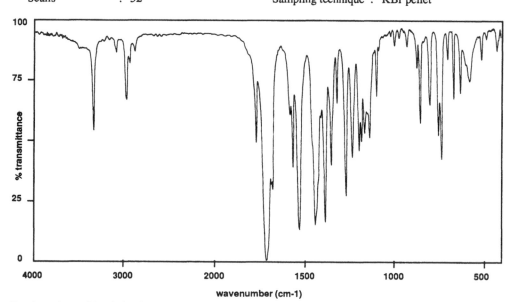

Chemical name	: 3-(3,5-dichlorophenyl)-N-isopropyl-2,4dioxoimidazolidine-1-carboxamide
Other names	: Glycophene, LFA 2043
Type	: fungicide
Brutoformula	: C13H13Cl2N3O3
Molecular mass	: 330.17286
Instrument	: Bruker IFS-85
Scans	: 32

CAS nr	: 35734-19-7
Exact mass	: 329.0333879
Optical resolution	: 2 cm-1
Sampling technique	: KBr pellet

Band maxima with relative intensity :

431	12	515	16	582	25	633	30	670	32
703	15	741	58	757	48	803	35	858	42
873	19	927	10	997	9	1101	31	1141	48
1169	46	1188	50	1202	54	1240	56	1278	73
1325	33	1359	60	1396	84	1452	85	1541	87
1574	60	1728	100	1780	50	2973	31	3084	11
3353	44								

238

COMPOUND : **Isobenzan**

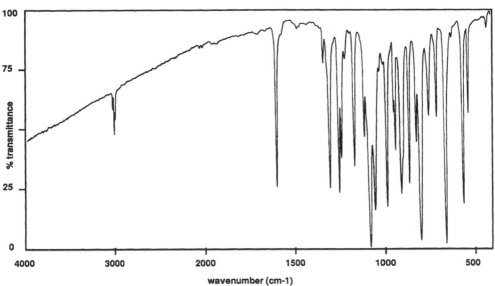

Chemical name	: 1,3,4,5,6,7,8,8-octachloro-1,3,3a,4,7,7a-hexahydro-4,7-methano-isobenzofuran		
Other names	: Telodrin		
Type	: insecticide		
Brutoformula	: C9H4Cl8O	CAS nr	: 297-78-9
Molecular mass	: 411.75563	Exact mass	: 407.7770373
Instrument	: Bruker IFS-85	Optical resolution	: 2 cm-1
Scans	: 32	Sampling technique	: KBr pellet

Band maxima with relative intensity :

534	43	565	81	663	98	711	45	755	44
803	97	826	55	866	73	913	77	943	58
951	42	989	82	1055	84	1081	100	1111	53
1167	65	1239	61	1252	76	1304	74	1337	21
1603	73	3010	50	3026	40				

COMPOUND : **Isocarbamid**

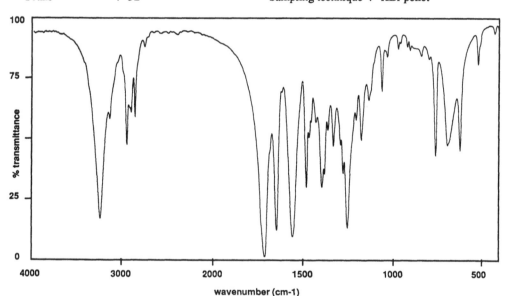

Chemical name	:	N-isobutyl-2-oxoimidazolidine-1-carboxamide		
Other names	:	Azolamide, Bayer 94871, MZ 166		
Type	:	herbicide		
Brutoformula	:	C8H15N3O2	CAS nr	: 30979-49-7
Molecular mass	:	185.22765	Exact mass	: 185.1164168
Instrument	:	Bruker IFS-85	Optical resolution	: 2 cm-1
Scans	:	32	Sampling technique	: KBr pellet

Band maxima with relative intensity :

521	19	626	55	696	53	764	57	972	12
1066	30	1140	34	1185	50	1264	87	1340	53
1404	70	1487	70	1565	91	1652	88	1720	100
2870	40	2959	52	3254	83				

COMPOUND : **Cyanuric acid**

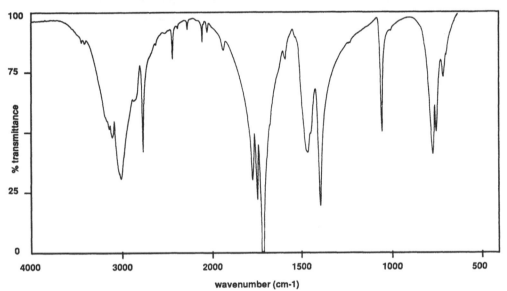

Chemical name   : 1,3,5-triazine-2,4,6(1H,3H,5H)-trione

Other names   : Isocyanuric acid
Type   : herbicide
Brutoformula   : C3H3N3O3   CAS nr   : 108-80-5
Molecular mass   : 129.07566   Exact mass   : 129.0174357
Instrument   : Bruker IFS-85   Optical resolution : 2 cm-1
Scans   : 32   Sampling technique : KBr pellet

Band maxima with relative intensity :

710	26	753	49	772	59	1053	49	1397	80
1463	58	1717	100	1752	77	1779	69	2108	10
2448	17	2780	56	3026	68	3117	50		

COMPOUND : **Isodrin**

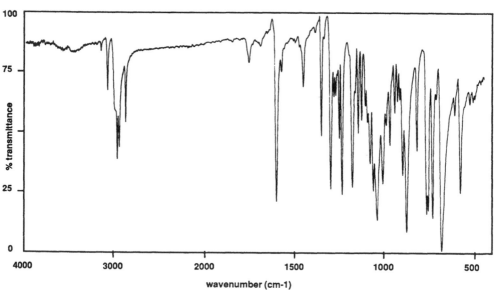

Chemical name	:	1,2,3,4,10,10-hexachloro-1,4,4b,5,8,8b-hexahydro-1,4:5,8-dimethano-naphtalene		
Other names	:	-		
Type	:	insecticide		
Brutoformula	:	C12H8Cl6	CAS nr	: 465-73-6
Molecular mass	:	364.91556	Exact mass	: 361.875154
Instrument	:	Bruker IFS-85	Optical resolution	: 2 cm-1
Scans	:	32	Sampling technique	: KBr pellet

Band maxima with relative intensity :

528	37	581	75	611	42	685	100	733	85
757	82	767	84	819	57	878	91	898	67
925	36	942	41	969	54	1006	71	1038	86
1059	74	1078	62	1106	38	1126	44	1145	49
1179	72	1237	75	1250	51	1270	33	1280	34
1300	73	1352	50	1451	29	1604	78	1756	19
2873	44	2945	54	2968	59	3074	30		

COMPOUND : **Isofenphos**

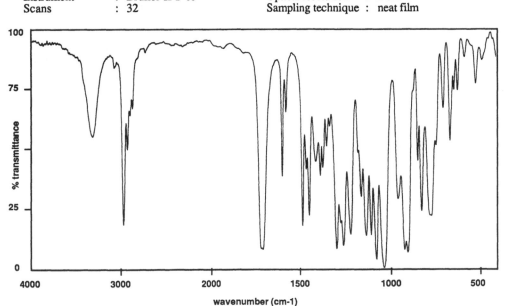

| Chemical name | : O-ethyl O-2-isopropoxycarbonylphenyl isopropylphosphoramidothioate |
| | |

Other names : Oftanol, Amaze, SRA 12869
Type : insecticide
Brutoformula : C15H24NO4PS
Molecular mass : 345.39659
Instrument : Bruker IFS-85
Scans : 32

CAS nr : 25311-71-1
Exact mass : 345.1163567
Optical resolution : 2 cm-1
Sampling technique : neat film

Band maxima with relative intensity :

522	21	586	10	624	24	645	24	667	46
705	32	772	77	829	75	849	54	909	93
961	70	1038	100	1081	96	1109	85	1136	86
1163	69	1220	85	1260	90	1298	91	1352	48
1373	57	1386	60	1412	54	1449	77	1487	82
1581	33	1603	60	1713	92	2934	49	2978	81
3317	44								

COMPOUND : **Isoproturon**

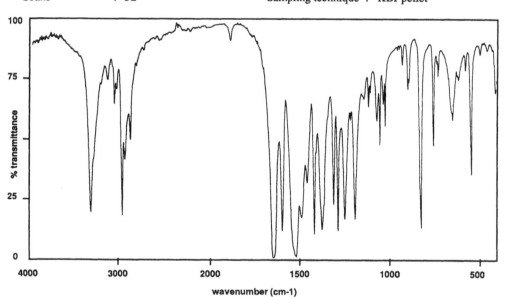

Chemical name   : 3-(4-isopropylphenyl-1,1-dimethylurea

Other names   : HOE 16410, CGA 18731, Arelon, Graminon
Type   : herbicide
Brutoformula   : C12H18N2O   CAS nr   : 34123-59-6
Molecular mass   : 206.29006   Exact mass   : 206.1419033
Instrument   : Bruker IFS-85   Optical resolution   : 2 cm-1
Scans   : 32   Sampling technique   : KBr pellet

Band maxima with relative intensity :

543	65	574	21	646	42	729	23	756	52
825	87	894	29	923	18	1018	44	1028	35
1050	52	1066	42	1112	36	1189	83	1245	83
1283	88	1307	77	1372	87	1415	89	1455	68
1518	99	1596	88	1649	100	1884	7	2864	49
2923	57	2954	81	3041	34	3113	23	3312	79

COMPOUND : **Lenacil**

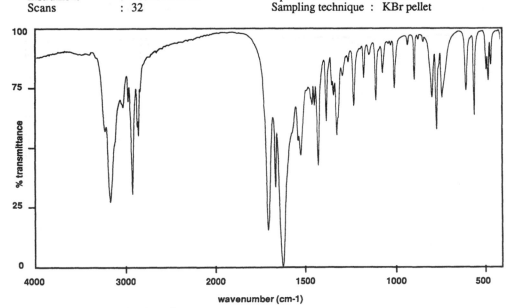

Chemical name   : 3-cyclohexyl-6,7-dihydro-1H-cyclopentapyrimidine-2,4-dione

Other names   : Venzar, Elbatan
Type   : herbicide
Brutoformula   : C13H18N2O2   CAS nr   : 2164-08-1
Molecular mass   : 234.30061   Exact mass   : 234.1368174
Instrument   : Bruker IFS-85   Optical resolution : 2 cm-1
Scans   : 32   Sampling technique : KBr pellet

Band maxima with relative intensity :

462	14	477	21	487	16	556	35	602	25
737	28	767	41	792	28	893	20	1002	24
1067	17	1106	29	1173	19	1229	31	1323	44
1339	27	1379	38	1424	57	1444	31	1458	31
1523	52	1625	100	1666	66	1707	84	2859	44
2929	69	2976	29	3172	72				

245

COMPOUND : **Lindane**

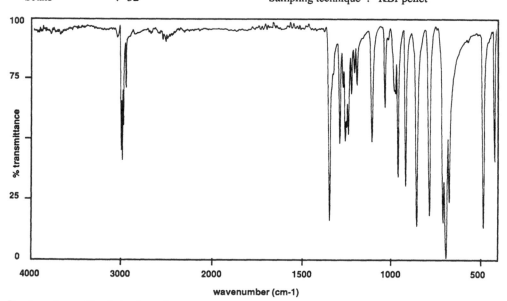

Gamma-HCH

Chemical name   : 1,2,3,4,5,6-hexachlorocyclohexane (gamma isomer)

Other names   : HCH-gamma, Hexachloran, Gammaexane
Type   : insecticide, rodenticide
Brutoformula   : C6H6Cl6   CAS nr   : 58-89-9
Molecular mass   : 290.83272   Exact mass   : 287.8600662
Instrument   : Bruker IFS-85   Optical resolution : 2 cm-1
Scans   : 32   Sampling technique : KBr pellet

Band maxima with relative intensity :

416	59	480	87	666	76	685	100	702	85
779	82	850	86	911	69	953	65	965	30
1024	36	1101	50	1183	26	1197	21	1215	30
1233	47	1251	50	1263	27	1282	51	1342	83
2494	7	2931	27	2961	43	2974	57	2987	53

COMPOUND : **Linuron**

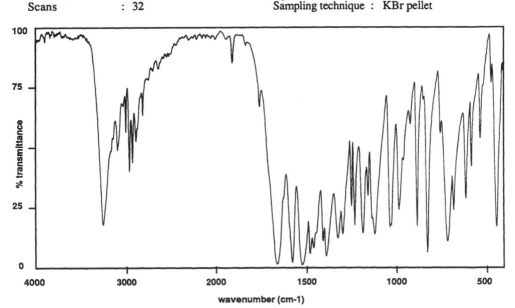

Chemical name	:	3- (3,4-dichlorophenyl)-1-methoxy-1-methylurea		

Other names	: Afalon, Methoxydiuron, Lorox, Linorox		
Type	: herbicide		
Brutoformula	: C9H10Cl2N2O2	CAS nr	: 330-55-2
Molecular mass	: 249.09825	Exact mass	: 248.0119268
Instrument	: Bruker IFS-85	Optical resolution	: 2 cm-1
Scans	: 32	Sampling technique	: KBr pellet

Band maxima with relative intensity :

440	84	467	22	529	46	577	58	610	72
676	76	711	90	748	44	823	94	882	83
917	40	979	76	1030	84	1114	87	1151	70
1181	86	1227	83	1244	75	1295	87	1322	88
1386	96	1405	90	1477	95	1522	100	1579	98
1663	99	1761	32	1908	14	2814	36	2893	47
2933	56	2969	60	3008	43	3103	51	3266	82

COMPOUND : **Malaoxon**

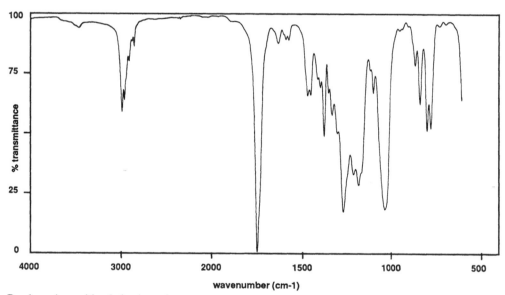

Chemical name   : [(dimethoxyphosphinyl)thio]-butanedioic acid, diethylester

Other names   : -
Type   : metabolite of Malathion
Brutoformula   : C10H19O7PS   CAS nr   : 1634-78-2
Molecular mass   : 314.2924   Exact mass   : 314.058902
Instrument   : Bruker IFS-85   Optical resolution : 2 cm-1
Scans   : 32   Sampling technique : neat film

Band maxima with relative intensity :

772	48	793	49	832	37	857	21	1028	82
1179	72	1265	83	1373	51	1464	34	1626	11
1741	100	2984	40						

COMPOUND : **Malathion**

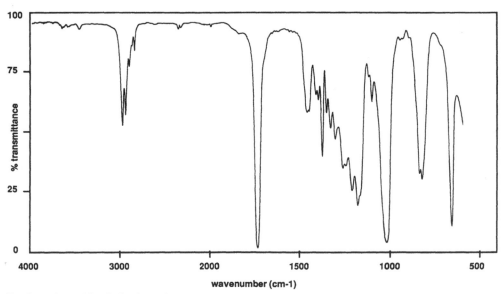

Chemical name : S-1,2-bis (ethoxycarbonyl) ethyl O,O-dimethylphosphorodithioate

Other names : Fyfanon, Sumitox, Carbofos
Type : insecticide, acaricide
Brutoformula : C10H19O6PS2   CAS nr : 121-75-5
Molecular mass : 330.35309   Exact mass : 330.0360614
Instrument : Bruker IFS-85   Optical resolution : 2 cm-1
Scans : 32   Sampling technique : neat film

Band maxima with relative intensity :

656	90	821	69	1018	97	1096	35	1176	81
1208	74	1299	52	1325	47	1349	40	1372	59
1393	34	1458	40	1736	100	2844	12	2948	41
2983	45								

COMPOUND : **Maleic hydrazide**

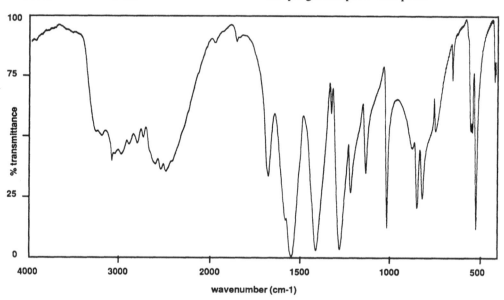

Chemical name   : 6-hydroxy-3(2H)pyridazinone

Other names   : Hydramin-C, Malzid, Burtolin
Type   : growth regulator
Brutoformula   : C4H4N2O2   CAS nr   : 122-33-1
Molecular mass   : 112.08868   Exact mass   : 112.027273
Instrument   : Bruker IFS-85   Optical resolution : 2 cm-1
Scans   : 32   Sampling technique : KBr pellet

Band maxima with relative intensity :

411	26	517	88	537	47	645	25	740	47
815	75	845	79	1007	88	1125	65	1210	73
1272	97	1318	39	1406	97	1543	100	1672	66
2468	63	3066	59						

COMPOUND : **Maneb**

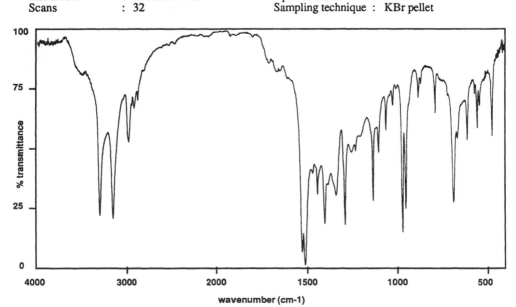

Chemical name   : manganese ethylene bis-dithiocarbamate

Other names   : Dithane, MEB, Manzate, Trimangol
Type   : fungicide
Brutoformula   : C4H6N2S4Mn   CAS nr   : 12427-38-2
Molecular mass   : 265.28382   Exact mass   : 264.8794397
Instrument   : Bruker IFS-85   Optical resolution   : 2 cm-1
Scans   : 32   Sampling technique : KBr pellet

Band maxima with relative intensity :

467	45	548	42	604	47	680	73	780	35
875	29	947	76	965	86	1016	32	1055	42
1096	52	1128	72	1287	83	1337	70	1400	82
1439	69	1509	100	1525	94	2975	47	3153	79
3296	78								

COMPOUND : **MCPA**

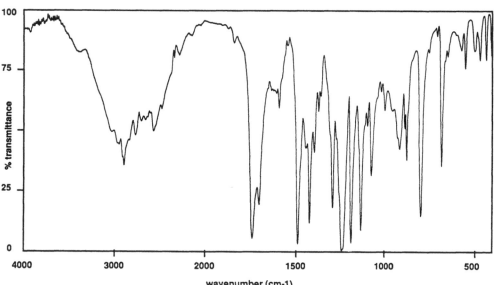

Chemical name   : (4-chloro-o-tolyloy) acetic acid

Other names   : Metaxon, Agroxone, Agritox, Chiptox
Type   : herbicide
Brutoformula   : C9H9ClO3   CAS nr   : 94-74-6
Molecular mass   : 200.62328   Exact mass   : 200.0240168
Instrument   : Bruker IFS-85   Optical resolution   : 2 cm-1
Scans   : 32   Sampling technique   : KBr pellet

Band maxima with relative intensity :

439	19	474	19	504	15	553	23	574	15
684	64	801	85	877	61	888	48	917	57
997	40	1077	68	1099	47	1136	91	1192	96
1246	100	1298	81	1378	41	1401	58	1428	88
1493	97	1597	39	1706	80	1744	94	2579	49
2781	51	2907	63						

COMPOUND : **MCPB**

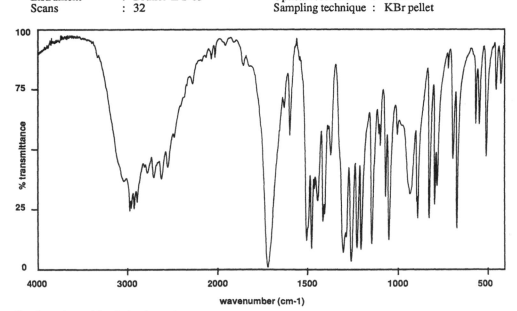

Chemical name	:	4-(4-chloro-0-tolyloxy) butyric acid

Other names	: Tropotox, Legumex, Trifolex				
Type	: herbicide				
Brutoformula	: C11H13ClO3		CAS nr	:	94-81-5
Molecular mass	: 228.67746		Exact mass	:	228.0553152
Instrument	: Bruker IFS-85		Optical resolution	:	2 cm-1
Scans	: 32		Sampling technique	:	KBr pellet

Band maxima with relative intensity :

421	21	447	24	500	52	539	38	558	38
660	83	683	53	713	15	771	65	784	73
813	78	880	78	923	68	994	43	1038	88
1056	69	1086	48	1095	43	1133	89	1192	92
1215	91	1246	97	1290	93	1364	52	1397	77
1406	80	1434	71	1455	66	1468	91	1495	88
1593	43	1625	32	1711	100	2552	58	2621	63
2705	62	2779	60	2884	72	2917	75	2966	76

COMPOUND : **Mecoprop**

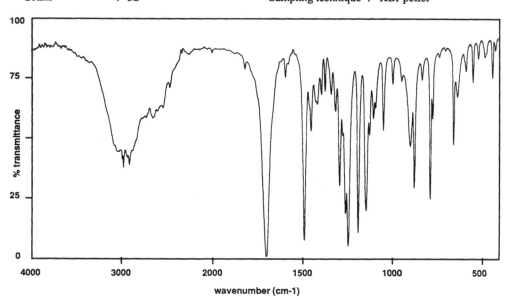

Chemical name   : ±-2-(4-chloro-o-tolyloxy) propionic acid

Other names   : MCPP, CMPP, Isocornox, Propal
Type   : herbicide
Brutoformula   : C10H11ClO3   CAS nr   : 7085-19-0
Molecular mass   : 214.65037   Exact mass   : 214.039666
Instrument   : Bruker IFS-85   Optical resolution : 4 cm-1
Scans   : 32   Sampling technique : KBr pellet

Band maxima with relative intensity :

444	24	485	15	522	16	554	26	593	21
640	32	661	52	779	41	792	75	838	25
880	71	902	53	996	27	1051	46	1106	41
1146	80	1190	89	1245	95	1261	81	1294	69
1316	38	1341	31	1376	29	1397	31	1419	35
1455	46	1492	93	1599	24	1704	100	2921	60
2987	62								

254

COMPOUND : **Mefluidide**

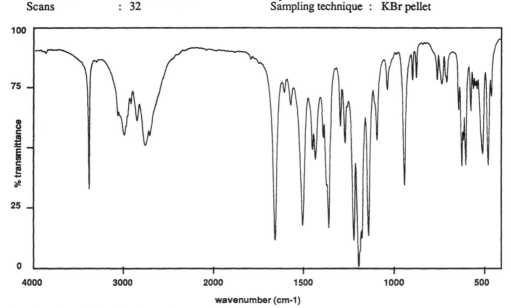

Chemical name   : 5'-(trifluoromethanesulphonamide) acet-2',4'-xylidide

Other names   : Methafluoridamid, Vistar, NC 18144, Embark
Type   : growth regulator, herbicide
Brutoformula   : C11H13F3N2O3S   CAS nr   : 53780-34-0
Molecular mass   : 310.29306   Exact mass   : 310.0598878
Instrument   : Bruker IFS-85   Optical resolution : 2 cm-1
Scans   : 32   Sampling technique : KBr pellet

Band maxima with relative intensity :

462	27	483	57	513	52	576	33	606	56
628	57	644	33	711	21	738	22	764	20
880	19	901	20	947	65	1039	24	1098	46
1146	86	1200	100	1226	88	1272	47	1298	40
1365	83	1395	44	1439	54	1456	50	1513	82
1579	31	1667	88	2769	48	2862	37	3007	43
3397	66								

COMPOUND : **Menazon**

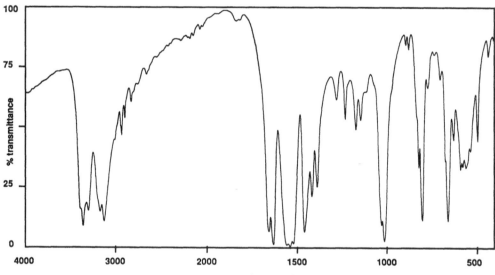

Chemical name	: O,O-dimethyl-S-(4,6-diamono-1,3,5-triazin-2-yl)-methylphosphorothioate
Other names	: Sayfos, PP 175
Type	: insecticide

Brutoformula	: C6H12N5O2PS2	CAS nr	: 78-57-9
Molecular mass	: 281.2886	Exact mass	: 281.0169988
Instrument	: Bruker IFS-85	Optical resolution	: 2 cm-1
Scans	: 32	Sampling technique	: KBr pellet

Band maxima with relative intensity :

435	20	496	56	590	68	628	55	662	89
807	89	824	67	877	17	1014	98	1146	46
1173	50	1234	46	1282	38	1391	74	1421	78
1463	93	1544	100	1637	99	1663	93	2838	38
2908	45	2943	52	3146	88	3376	90		

COMPOUND : **Metalaxyl**

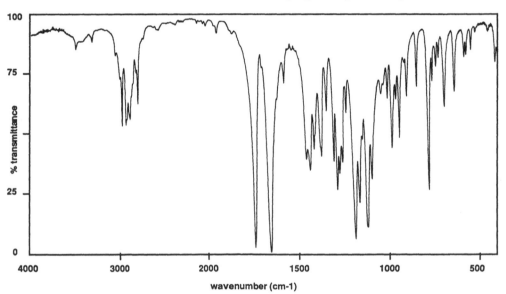

Chemical name   : methyl-N-(2-methoxyacetyl)-N-(2,6-xylyl)-DL-alaninate

Other names   : Metaxanine, CGA 48988, Ridomil
Type   : fungicide
Brutoformula   : C15H21NO4   CAS nr   : 57837-19-1
Molecular mass   : 279.33892   Exact mass   : 279.1470462
Instrument   : Bruker IFS-85   Optical resolution : 2 cm-1
Scans   : 32   Sampling technique : KBr pellet

Band maxima with relative intensity :

784	74	949	52	991	56	1104	69	1127	90
1173	79	1198	94	1267	62	1284	66	1298	74
1316	61	1389	59	1427	56	1449	65	1670	100
1759	97	2996	45						

257

COMPOUND : **Metham (sodium salt)**

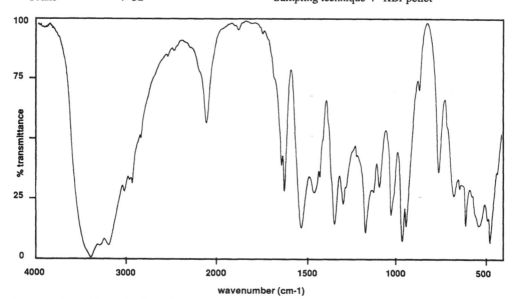

Chemical name : sodium dimethyldithiocarbamate

Other names : Vapam, Monam, N 869, Trimaton
Type : fungicide, herbicide, insecticide
Brutoformula : C2H4NS2Na   CAS nr   : 137-42-8
Molecular mass : 129.17065   Exact mass   : 128.968298
Instrument : Bruker IFS-85   Optical resolution : 2 cm-1
Scans : 32   Sampling technique : KBr pellet

Band maxima with relative intensity :

468	93	531	86	603	86	670	73	756	63
932	86	952	92	1017	81	1085	69	1163	89
1292	77	1341	85	1459	72	1528	87	1625	71
2109	43	3379	100						

COMPOUND : **Metamitron**

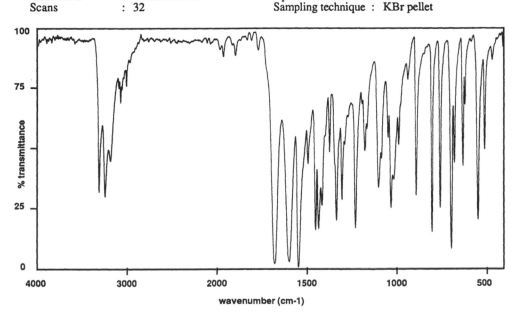

| Chemical name | : 4-amino-4,5-dihydro-3-methyl-6-phenyl-1,2,4-triazin-5-one |
| | |

Other names : DRW 1139, Goltix, Bayer 134028, Herbrak
Type : herbicide
Brutoformula : C10H10N4O   CAS nr : 41394-05-2
Molecular mass : 202.2174   Exact mass : 202.0854529
Instrument : Bruker IFS-85   Optical resolution : 2 cm-1
Scans : 32   Sampling technique : KBr pellet

Band maxima with relative intensity :

507	49	543	79	615	30	626	56	676	55
693	92	755	74	800	84	889	69	934	20
982	47	1025	74	1041	44	1094	66	1172	50
1225	83	1300	71	1331	80	1371	50	1413	73
1433	83	1449	84	1491	56	1544	100	1597	97
1678	98	3010	23	3074	30	3245	70	3310	68

COMPOUND : **Metazachlor**

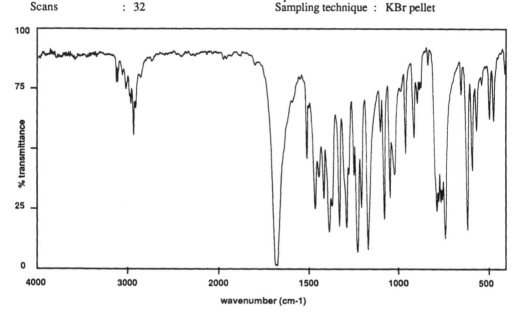

Chemical name   : 2-chloro-N-(2,6-dimethylphenyl)-N-(pyrazolylmethyl) acetamide

Other names   : Butisan S, BAS 479H
Type   : herbicide
Brutoformula   : C14H16ClN3O   CAS nr   : 67129-08-2
Molecular mass   : 277.75612   Exact mass   : 277.0981804
Instrument   : Bruker IFS-85   Optical resolution : 2 cm-1
Scans   : 32   Sampling technique : KBr pellet

Band maxima with relative intensity :

409	19	475	39	497	38	568	43	594	60
622	85	652	28	746	88	769	73	792	77
837	15	897	31	917	46	964	53	1025	61
1051	71	1084	80	1103	43	1176	93	1212	75
1235	94	1253	61	1296	84	1336	83	1394	85
1422	71	1448	62	1471	75	1515	54	1686	100
2945	43	2973	30	3024	24	3120	21	3135	21

COMPOUND : **Methabenzthiazuron**

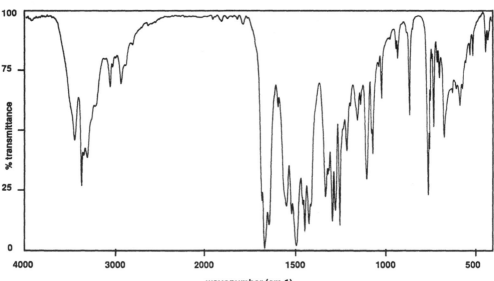

Chemical name   : 1-(benzothiazol-2-yl)-1,3-dimethylurea

Other names	: Tribunil, Bayer 74283		
Type	: herbicide		
Brutoformula	: C10H11N3OS	CAS nr	: 18691-97-9
Molecular mass	: 221.27867	Exact mass	: 221.062277
Instrument	: Bruker IFS-85	Optical resolution	: 2 cm-1
Scans	: 32	Sampling technique	: KBr pellet

Band maxima with relative intensity :

437	16	509	18	582	39	671	53	696	27
729	48	753	58	761	77	863	43	927	19
1015	36	1067	60	1103	70	1154	45	1214	58
1256	90	1279	84	1297	88	1334	78	1425	89
1448	92	1494	98	1549	82	1647	89	1672	100
2941	29	3060	31	3376	73	3448	53		

261

COMPOUND : **Methamidophos**

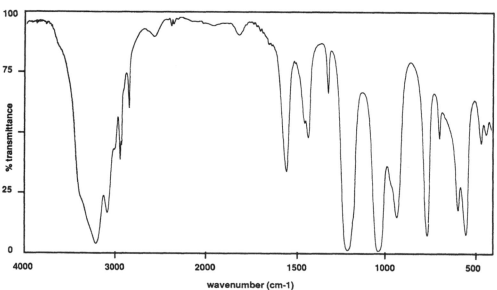

Chemical name   : O,S-dimethylphosporamidothioate

Other names   : Tamaron, Monitor, SRA 71628
Type   : acaricide, insecticide
Brutoformula   : C2H8NO2PS
Molecular mass   : 141.12532
Instrument   : Bruker IFS-85
Scans   : 32

CAS nr   : 10265-92-6
Exact mass   : 141.0013349
Optical resolution : 2 cm-1
Sampling technique : neat film

Band maxima with relative intensity :

471	55	559	93	600	83	700	52	774	93
939	85	1044	100	1220	99	1321	32	1437	51
1562	65	1812	7	2846	38	2949	59	3105	82
3238	95								

COMPOUND : **Methidathion**

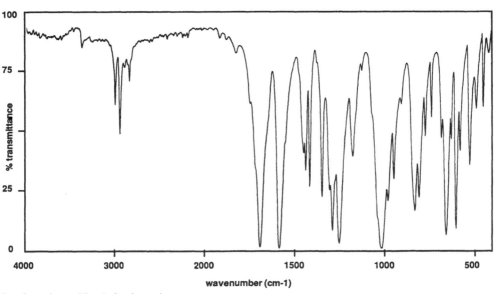

| Chemical name | : S-2,3-dihydro-5-methoxy-2-oxo-1,3,4-thiadiazol-3-ylmethyl O,O- |
| dimethyl | phosphorodithioate |

Other names : Supracid, Ultracid, Somonil, GS 13005
Type : acaricide, insecticide

Brutoformula	: C6H11N2O4PS3	CAS nr	: 950-37-8
Molecular mass	: 302.31933	Exact mass	: 301.9618555
Instrument	: Bruker IFS-85	Optical resolution	: 2 cm-1
Scans	: 32	Sampling technique	: KBr pellet

Band maxima with relative intensity :

421	16	450	39	489	40	524	64	578	58
602	91	628	53	658	94	683	53	738	44
773	52	805	78	829	84	945	70	1012	100
1174	60	1248	97	1288	92	1346	78	1414	73
1437	66	1583	99	1692	99	2839	28	2945	51
2999	39								

COMPOUND : **Methiocarb**

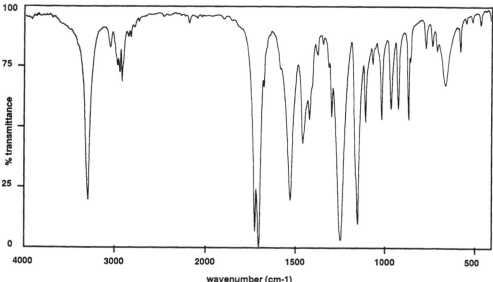

Chemical name   : (3,5-dimethyl-4-methylthio-phenyl)-N-methyl carbamate

Other names   : H321, Mercaptodimethur
Type   : insecticide, acaricide, molluscidide
Brutoformula   : C11H15NO2S   CAS nr   : 2032-65-7
Molecular mass : 225.3077   Exact mass   : 225.0823431
Instrument   : Bruker IFS-85   Optical resolution : 2 cm-1
Scans   : 32   Sampling technique : KBr pellet

Band maxima with relative intensity :

579	17	664	32	733	15	771	16	870	46
927	42	967	42	1017	46	1067	22	1109	47
1156	90	1251	97	1298	45	1423	46	1461	56
1533	80	1709	100	1731	93	2922	29	2947	25
3052	15	3310	79						

COMPOUND : **Methomyl**

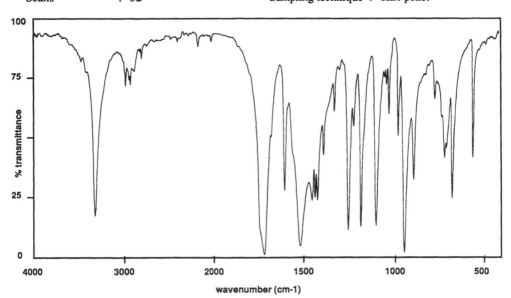

Chemical name   : S-methyl-N-(methylcarbamoyloxy) thioacetimidate

Other names   : Lannate, Nudrin
Type   : acaricide, insecticide
Brutoformula   : C5H10N2O2S   CAS nr   : 16752-77-5
Molecular mass   : 162.20765   Exact mass   : 162.0462933
Instrument   : Bruker IFS-85   Optical resolution : 2 cm-1
Scans   : 32   Sampling technique : KBr pellet

Band maxima with relative intensity :

552	58	664	75	708	58	764	33	884	68
935	98	973	49	1025	40	1038	27	1094	87
1180	87	1220	45	1247	89	1325	38	1382	57
1414	76	1428	75	1444	76	1508	96	1598	72
1708	100	2180	11	2928	27	2980	28	3300	83

COMPOUND : **Methoprene**

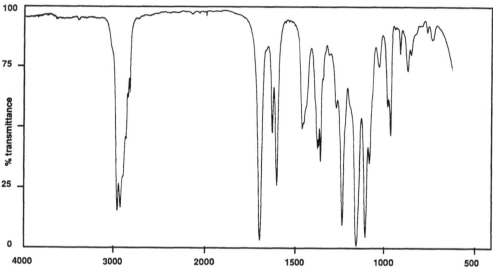

Chemical name	:	isopropyl (E,E)-11-methoxy-3,7,11-trimethyl-2,4-dodecadienoate

Other names	: Altosid, ZR 515, Precor, Apex		
Type	: insecticide		
Brutoformula	: C19H34O3	CAS nr	: 40596-69-8
Molecular mass	: 310.48103	Exact mass	: 310.2507787
Instrument	: Bruker IFS-85	Optical resolution	: 2 cm-1
Scans	: 32	Sampling technique	: neat film

Band maxima with relative intensity :

738	12	872	25	914	18	966	53	1028	23
1086	64	1110	96	1160	100	1239	91	1362	64
1379	58	1465	50	1612	74	1637	51	1708	97
2824	34	2937	83	2974	84				

266

COMPOUND : **Methoprotryne**

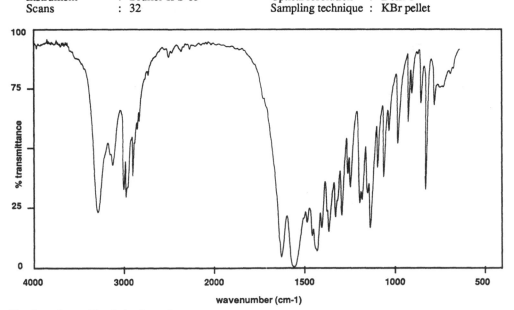

Chemical name   : 2-isopropylamino-4(3-methoxypropylamino)-6-methylthio-1,3,5-triazine

Other names   : Gesaran, G 36393
Type   : herbicide
Brutoformula   : C11H21N5OS   CAS nr   : 841-06-5
Molecular mass   : 271.38292   Exact mass   : 271.1466694
Instrument   : Bruker IFS-85   Optical resolution : 2 cm-1
Scans   : 32   Sampling technique : KBr pellet

Band maxima with relative intensity :

767	30	808	66	841	29	894	25	912	37
968	47	1017	41	1043	61	1079	57	1115	83
1174	72	1227	65	1243	60	1273	77	1308	78
1344	84	1383	83	1406	93	1532	100	1605	95
2870	62	2943	71	2970	67	3097	57	3255	77

COMPOUND : **Methoxychlor**

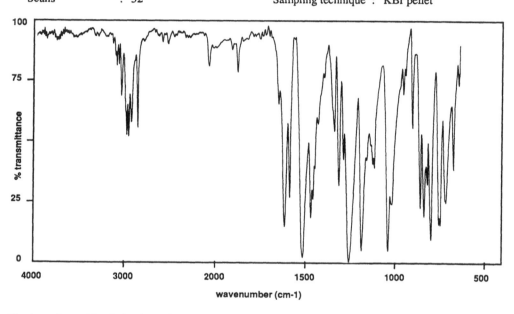

Chemical name : 1,1,1-trichloro-2,2-bis (4-methoxyphenyl) ethane

Other names : DMDT, Marlate
Type : acaricide, herbicide
Brutoformula : C16H15Cl3O2   CAS nr : 72-43-5
Molecular mass : 345.65575   Exact mass : 344.0137565
Instrument : Bruker IFS-85   Optical resolution : 2 cm-1
Scans : 32   Sampling technique : KBr pellet

Band maxima with relative intensity :

632	25	664	61	708	75	740	84	793	90
812	67	832	81	852	77	892	43	941	29
1032	95	1109	60	1183	95	1253	100	1280	56
1307	67	1330	44	1463	81	1510	98	1582	72
1612	84	1641	33	1866	19	2835	42	2903	40
2934	46	2955	45	3011	29	3037	15	3062	16

COMPOUND : **Methoxyphenone**

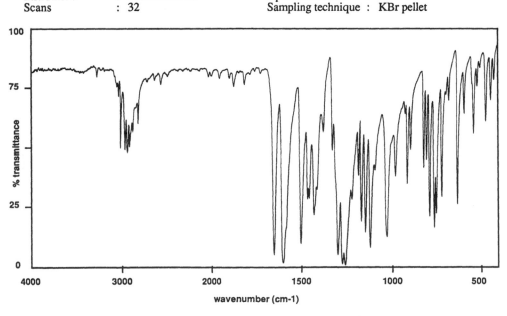

Chemical name   : 3, 3'-dimethyl-4-methoxy-benzophenone

Other names   : NK 049
Type   :
Brutoformula   : C16H16O2   CAS nr   :
Molecular mass   : 240.3047   Exact mass   : 240.1150218
Instrument   : Bruker IFS-85   Optical resolution : 2 cm-1
Scans   : 32   Sampling technique : KBr pellet

Band maxima with relative intensity :

428	21	445	29	472	38	521	23	540	44
593	35	629	74	679	29	717	70	747	77
758	83	784	79	802	55	818	58	894	50
912	65	977	62	1025	87	1119	92	1145	85
1168	81	1184	61	1254	100	1295	95	1325	51
1377	43	1427	78	1462	71	1499	90	1597	99
1650	95	2826	39	2920	49	2942	51	2970	50
3018	50	3043	27						

269

COMPOUND : **Metiram**

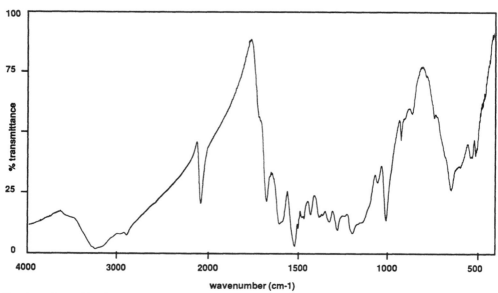

Chemical name	: tris[amine[ethylenebis(dithiocarbamato)] zinc(2+)] tetrahydro-1,2,4,7-dithiadiazocine-3,8-dithione polymer
Other names	: Polyram-Combi, Carbatene
Type	: fungicide
Brutoformula	: 1088.61511
Molecular mass	: C16H33N11S16Zn3
Instrument	: Bruker IFS-85
Scans	: 32

CAS nr	: 9006-42-2
Exact mass	: 1082.6326032
Optical resolution	: 2 cm-1
Sampling technique	: KBr pellet

Band maxima with relative intensity :

508	61	536	61	643	75	919	54	1001	88
1193	93	1275	92	1428	85	1519	99	1605	89
1678	80	2078	80	3243	100				

COMPOUND : **Metobromuron**

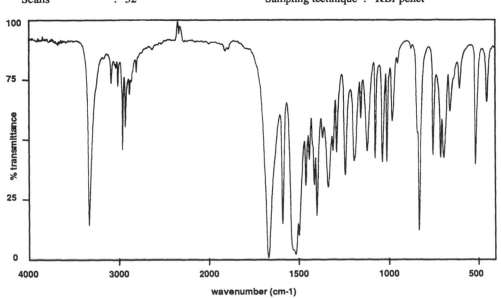

Chemical name	: 3-(4-bromophenyl-1-methoxy-1-methylurea		
Other names	: Patoran, CIBA 3126, Pattonex		
Type	: herbicide		
Brutoformula	: C9H11BrN2O2	CAS nr	: 3060-89-7
Molecular mass	: 259.10422	Exact mass	: 258.0004352
Instrument	: Bruker IFS-85	Optical resolution	: 2 cm-1
Scans	: 32	Sampling technique	: KBr pellet

Band maxima with relative intensity :

448	33	510	60	598	27	652	37	685	57
703	57	745	56	823	88	977	41	1005	59
1030	59	1070	57	1114	54	1152	40	1188	58
1237	64	1286	55	1305	54	1331	70	1394	82
1409	69	1437	58	1456	69	1509	98	1586	85
1664	100	2811	21	2889	30	2931	44	2962	54
3017	27	3090	26	3327	86				

271

COMPOUND : **2,4-DB**

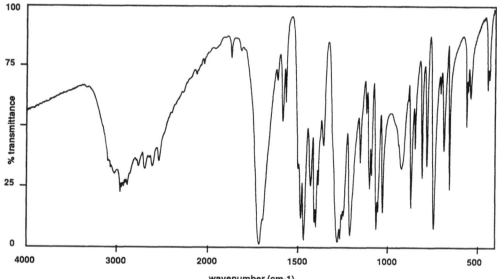

Chemical name   : 4-(2,4-dichlorophenoxy) butyric acid

Other names   : Butirex, Butormone, Butoxone
Type   : herbicide
Brutoformula   : C10H10Cl2O3   CAS nr   : 94-82-6
Molecular mass   : 249.095   Exact mass   : 248.000694
Instrument   : Bruker IFS-85   Optical resolution : 2 cm-1
Scans   : 32   Sampling technique : KBr pellet

Band maxima with relative intensity :

442	34	536	37	550	34	558	49	653	76
683	60	702	35	740	93	778	66	806	71
845	59	868	84	919	67	1026	86	1053	86
1060	93	1088	70	1100	76	1119	44	1154	65
1211	96	1267	97	1281	98	1358	58	1389	74
1402	92	1410	90	1431	75	1468	98	1485	88
1569	39	1586	47	1615	30	1716	100	2547	65
2618	67	2699	68	2772	68	2971	78		

272

COMPOUND : **Metolachlor**

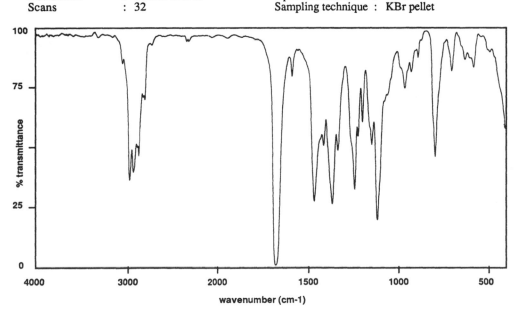

Chemical name   :   2-chloro-6'-ethyl-N-(2-methoxy-1-methylethyl) acet-o-toluidide

Other names   :   Dual, CGA 24705
Type   :   herbicide
Brutoformula   :   C15H22ClNO2   CAS nr   :   51218-45-2
Molecular mass   :   283.80109   Exact mass   :   283.1338957
Instrument   :   Bruker IFS-85   Optical resolution   :   2 cm-1
Scans   :   32   Sampling technique   :   KBr pellet

Band maxima with relative intensity :

581	15	700	17	787	53	960	24	1112	80
1146	48	1199	38	1241	67	1333	50	1363	73
1461	72	1590	19	1674	100	2935	60	2976	63

273

COMPOUND : **Metomeclan**

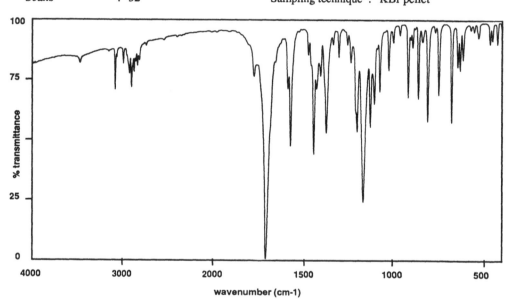

| Chemical name | : 1-(3,5-dichlorophenyl)-3-methoxymethyl-pyrolidin-2,5-dion |

Other names : CO 6054
Type : herbicide
Brutoformula : C12H11Cl2NO3
Molecular mass : 288.13237
Instrument : Bruker IFS-85
Scans : 32

CAS nr : -
Exact mass : 287.0115923
Optical resolution : 2 cm-1
Sampling technique : KBr pellet

Band maxima with relative intensity :

430	9	469	9	531	6	613	15	627	20
640	18	673	42	746	30	806	41	857	32
887	10	912	31	1017	20	1068	29	1101	34
1124	44	1163	75	1199	46	1235	16	1304	14
1376	46	1407	22	1445	55	1476	13	1576	52
1592	28	1710	100	2873	20	2901	27	2925	21
2992	17	3087	28						

COMPOUND : **Metoxuron**

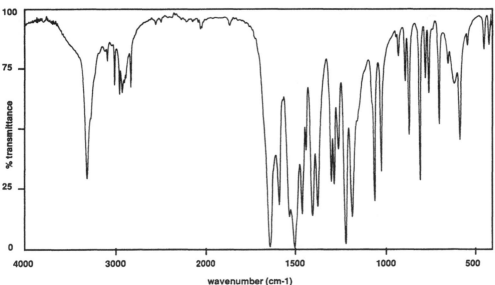

Chemical name   :   3-(3-chloro-4-methoxyphenyl)-1,1-dimethylurea

Other names   :   Dosanex, SAN 171H, Purivel, Deftor
Type   :   herbicide
Brutoformula   :   C10H13ClN2O2   CAS nr   :   19937-59-8
Molecular mass   :   228.68031   Exact mass   :   228.0665475
Instrument   :   Bruker IFS-85   Optical resolution   :   2 cm-1
Scans   :   32   Sampling technique :   KBr pellet

Band maxima with relative intensity :

416	14	445	16	583	55	646	22	697	48
754	34	772	28	804	71	866	52	887	29
923	19	1023	67	1061	80	1188	87	1223	98
1259	58	1283	73	1298	72	1375	82	1405	86
1437	58	1461	85	1505	100	1591	81	1645	99
2833	31	2929	33	2957	34	3010	30	3088	19
3314	69								

COMPOUND : **Metribuzin**

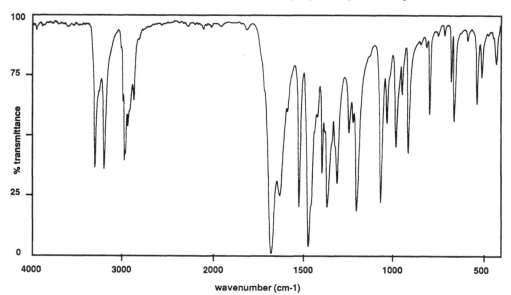

Chemical name	:	4-amino-6-tert.butyl-4,5-dihydro-3-methylthio-1,2,4-triazin-5-one

Other names	:	Sencor, DIC 1468, Bayer94337, Lexone			
Type	:	herbicide			
Brutoformula	:	C8H14N4OS	CAS nr	:	21087-64-9
Molecular mass	:	214.28698	Exact mass	:	214.088824
Instrument	:	Bruker IFS-85	Optical resolution	:	2 cm-1
Scans	:	32	Sampling technique	:	KBr pellet

Band maxima with relative intensity :

431	19	509	24	535	36	658	43	676	26
793	40	908	57	942	32	973	54	1024	44
1057	78	1191	81	1237	48	1303	70	1359	80
1389	65	1463	97	1518	80	1629	75	1675	100
2867	35	2969	60	3201	64	3304	63		

COMPOUND : **Metsulfuron-methyl**

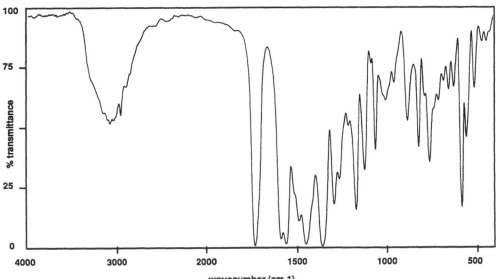

Chemical name	: methyl-2-[[[[(4-methoxy-6-methyl-1,3,5-triazin-2-yl) amino] carbonyl] amino] sulfonyl] benzoate

Other names : DPX-T 6376
Type : herbicide

Brutoformula	: C14H14N5O6S	CAS nr	: 74223-64-6
Molecular mass	: 381.36555	Exact mass	: 381.0742923
Instrument	: Bruker IFS-85	Optical resolution	: 4 cm-1
Scans	: 32	Sampling technique	: KBr pellet

Band maxima with relative intensity :

472	13	513	33	557	54	584	83	624	32
653	33	759	64	817	58	881	47	1001	38
1058	59	1120	67	1170	84	1294	82	1359	100
1450	99	1562	98	1733	99	3070	47		

COMPOUND : **Mevinphos (cis/trans mixture)**

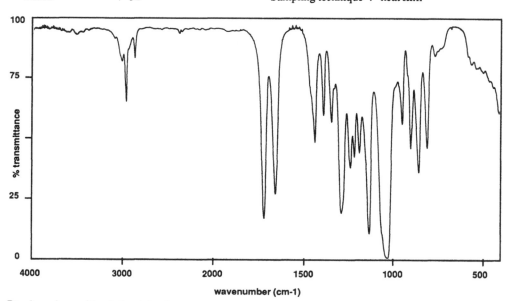

| Chemical name | : 2-methoxycarbonyl-1-methylvinyl dimethyl phosphate |
| | |

Other names   : Phosdrin, OS 2046, Apavinphos
Type   : acaricide, insecticide

Brutoformula	: C7H13O6P	CAS nr	: 7786-34-7
Molecular mass	: 224.15182	Exact mass	: 224.0449684
Instrument	: Bruker IFS-85	Optical resolution	: 2 cm-1
Scans	: 32	Sampling technique	: neat film

Band maxima with relative intensity :

808	53	856	63	899	53	946	43	1031	100
1136	89	1189	55	1217	56	1242	61	1293	80
1345	42	1389	39	1438	50	1660	72	1724	82
2858	14	2959	32						

COMPOUND : **Monocrotophos**

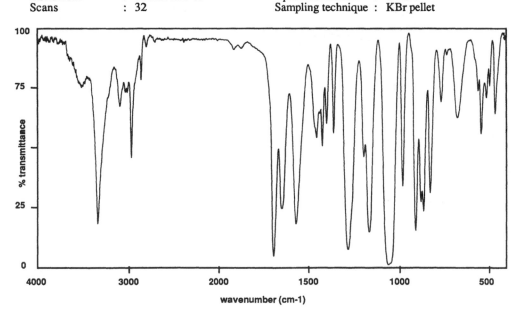

Chemical name : dimethyl-(E)-1-methyl-2-methylcarbamoylvinyl phosphate

Other names   : Nuvacron, Azodrin, Bilobran, C1414
Type   : acaricide, insecticide
Brutoformula   : C7H14NO5P   CAS nr   : 6923-22-4
Molecular mass : 223.16709   Exact mass   : 223.0609521
Instrument   : Bruker IFS-85   Optical resolution : 2 cm-1
Scans   : 32   Sampling technique : KBr pellet

Band maxima with relative intensity :

459	35	489	24	507	29	536	44	668	37
758	30	815	69	853	77	898	85	972	66
1050	100	1157	86	1189	54	1273	93	1354	44
1392	40	1416	49	1447	46	1559	82	1638	76
1685	96	2857	21	2959	54	3088	32	3319	82

COMPOUND : **Monolinuron**

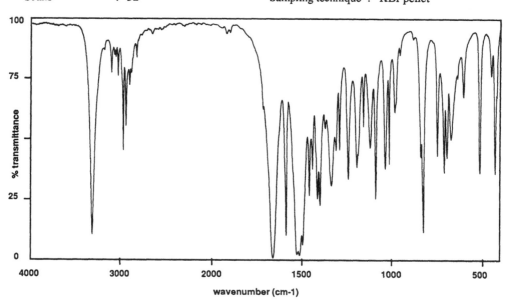

Chemical name   : 3-(4-chlorophenyl)-1-methoxy-1-methylurea

Other names   : Aresin, HOE 2747
Type   : herbicide
Brutoformula   : C9H11ClN2O2   CAS nr   : 1746-81-2
Molecular mass   : 214.65322   Exact mass   : 214.0508983
Instrument   : Bruker IFS-85   Optical resolution   : 2 cm-1
Scans   : 32   Sampling technique   : KBr pellet

Band maxima with relative intensity :

430	64	449	23	514	64	603	32	671	50
693	57	708	64	746	57	825	89	837	57
978	38	1009	60	1031	62	1086	75	1116	53
1153	44	1190	61	1238	66	1286	54	1305	54
1332	69	1397	77	1411	74	1438	62	1457	73
1513	98	1589	90	1664	100	2812	14	2890	26
2933	43	2964	54	3021	22	3095	21	3322	89

COMPOUND : **Monuron**

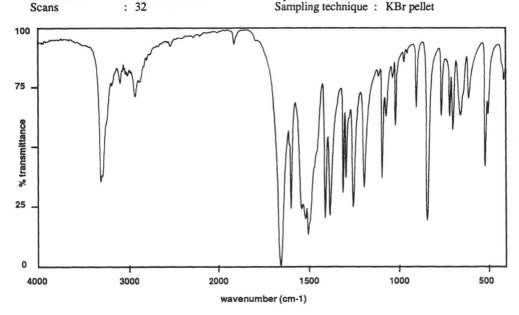

Chemical name   : 3-(4-chlorophenyl)-1,1-dimethylurea

Other names   : CMU, Telvar, Monurex
Type   : herbicide
Brutoformula   : C9H11ClN2O   CAS nr   : 150-68-5
Molecular mass   : 198.65382   Exact mass   : 198.0559842
Instrument   : Bruker IFS-85   Optical resolution : 2 cm-1
Scans   : 32   Sampling technique : KBr pellet

Band maxima with relative intensity :

412	21	512	57	604	28	649	36	692	42
710	36	755	36	831	80	897	32	1011	40
1065	36	1087	62	1188	66	1245	75	1287	62
1303	69	1375	78	1402	79	1494	86	1591	75
1645	100	1912	6	2931	28	3097	22	3298	64

COMPOUND : **Myclobutanil**

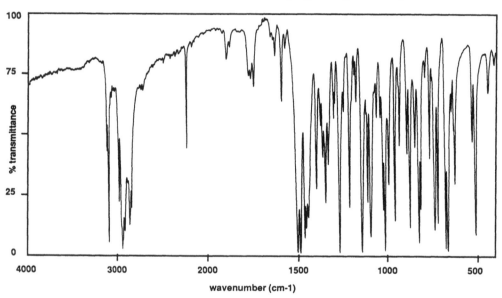

Chemical name : 2-butyl-2-(4-chlorophenyl)-3-(1,2,4-triazol-1-yl)-1-propanitril

Other names : Systhane, RH 3866, WJZ 2584A
Type : fungicide
Brutoformula : C15H17ClN4   CAS nr   : 88671-89-0
Molecular mass : 288.8   Exact mass   :
Instrument : Bruker IFS-85   Optical resolution   : 2 cm-1
Scans : 32   Sampling technique : KBr pellet

Band maxima with relative intensity :

514	93	530	53	625	71	666	99	678	98
720	87	738	90	765	60	815	81	826	96
846	55	876	89	892	58	932	55	959	86
991	71	1014	99	1023	82	1098	93	1113	78
1147	99	1217	80	1275	99	1337	62	1352	78
1368	59	1404	72	1449	85	1461	89	1469	93
1493	100	1509	99	2855	79	2875	87	2930	89
2957	97	2993	77	3113	94	3128	55		

282

COMPOUND : **Nabam**

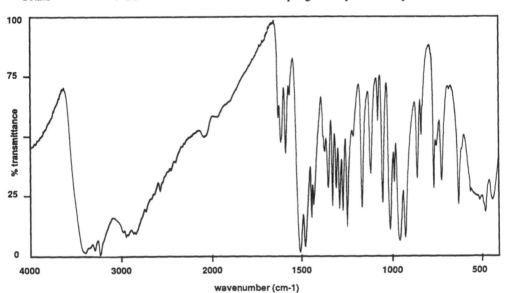

Chemical name	: disodium ethylenebis(dithiocarbamate)

Other names : Dithane D14, DSE, Parzate, Nabasan
Type : fungicide
Brutoformula : C4H6N2S4Na   CAS nr   : 88671-89-0
Molecular mass : 256.32536   Exact mass   : 255.9209468
Instrument : Bruker IFS-85   Optical resolution : 2 cm-1
Scans : 32   Sampling technique : KBr pellet

Band maxima with relative intensity :

438	77	478	82	624	79	720	68	761	72
834	49	856	67	924	92	956	94	983	69
1006	89	1046	78	1072	43	1109	65	1159	80
1241	88	1265	81	1284	80	1302	72	1324	79
1348	71	1367	56	1429	78	1440	84	1476	96
1505	99	1585	57	1613	52	1627	42	2947	92
3229	100	3396	99						

COMPOUND : **Naled**

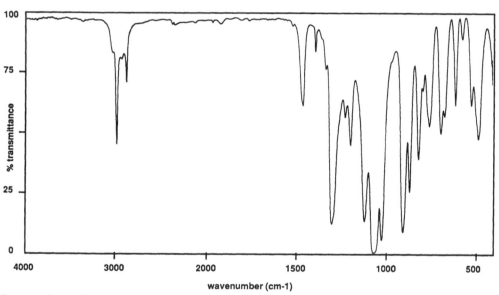

Chemical name	: 1,2-dibromo-2,2-dichloroethyl dimethylphosphate		
Other names	: Dibrom, Naurycid, Bromex, Flibol		
Type	: acaricide, insecticide		
Brutoformula	: C4H7Br2Cl2O4P	CAS nr	: 300-76-5
Molecular mass	: 380.78575	Exact mass	: 377.7826788
Instrument	: Bruker IFS-85	Optical resolution	: 2 cm-1
Scans	: 32	Sampling technique	: neat film

Band maxima with relative intensity :

477	51	515	37	568	9	604	37	685	49
750	46	810	60	858	74	893	91	1012	94
1058	100	1109	86	1188	54	1217	42	1291	87
1383	14	1452	37	2857	27	2960	54		

284

COMPOUND : **(2-Naphtyloxy) acetic acid**

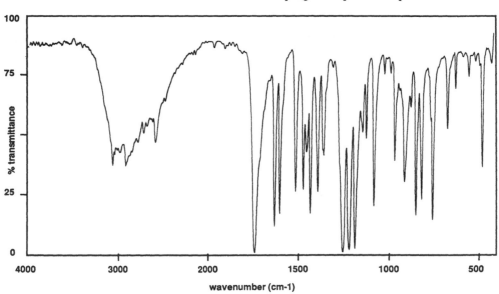

Chemical name : 2-naphtalenyloxy acetic acid

Other names : Naphtoxy acetic acid, BNOA
Type : growth regulator
Brutoformula : C12H10O3
Molecular mass : 202.2117
Instrument : Bruker IFS-85
Scans : 32

CAS nr : 120-23-0
Exact mass : 202.0629883
Optical resolution : 2 cm-1
Sampling technique : KBr pellet

Band maxima with relative intensity :

425	19	475	63	547	24	622	29	667	47
749	85	810	77	843	83	909	69	963	60
986	23	1021	23	1078	80	1121	51	1141	48
1183	98	1215	98	1250	99	1357	58	1389	74
1430	83	1451	57	1468	73	1510	73	1599	83
1628	88	1739	100	2587	53	2913	63	3059	63

COMPOUND : **1-Naphtyl-acetamide**

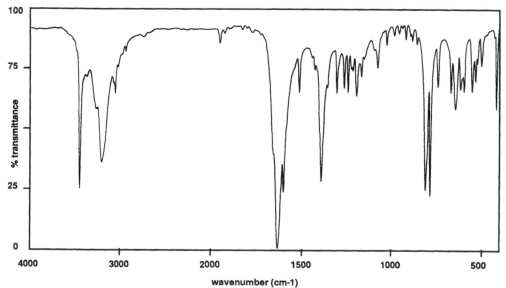

Chemical name : 1-naphtalenyl acetamide

Other names : -
Type : growth regulator
Brutoformula : C12H11NO
Molecular mass : 185.22757
Instrument : Bruker IFS-85
Scans : 32

CAS nr : 31093-43-3
Exact mass : 185.0840579
Optical resolution : 2 cm-1
Sampling technique : KBr pellet

Band maxima with relative intensity :

416	40	495	21	526	28	544	32	589	32
608	31	635	40	660	33	733	30	775	
801	74	912	10	1018	13	1070	22	1161	26
1189	35	1211	24	1238	33	1260	31	1301	33
1386	71	1510	33	1595	75	1625	100	1947	12
3046	34	3195	63	3434	74				

COMPOUND   : **1-Naphthol**

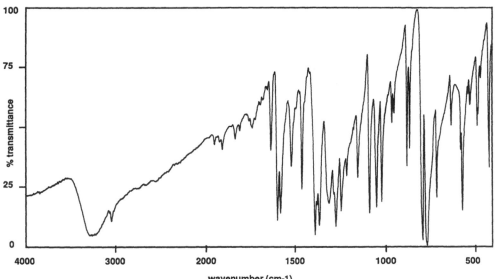

Chemical name   : 1-naphthol

Other names   : -
Type   : metabolite of 1-naphthylacetamide
Brutoformula   : C10H8O   CAS nr   : -
Molecular mass   : 144.1746   Exact mass   : 144.057510
Instrument   : Bruker IFS-85   Optical resolution : 2 cm-1
Scans   : 32   Sampling technique : KBr pellet

Band maxima with relative intensity :

415	67	462	29	483	49	522	40	567	85
574	59	628	49	709	79	764	100	789	97
860	58	876	66	946	43	959	47	1015	81
1043	83	1083	86	1147	71	1208	70	1240	84
1270	91	1308	81	1362	90	1386	94	1457	75
1516	65	1580	85	1598	88	1633	58	1737	49
1833	54	1907	58	1951	56	3050	88	3265	94

COMPOUND : **1-Naphthyl acetic acid**

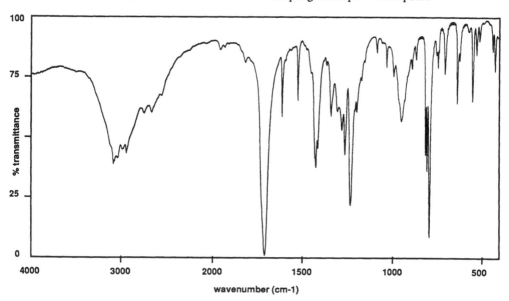

Chemical name : 1-naphtalenylacetic acid

Other names : Naphthyl acetic acid, Rhizopon-B, Fruitofix
Type : growth regulator
Brutoformula : C12H10O2   CAS nr   : 86-87-3
Molecular mass : 186.2123   Exact mass   : 186.0680742
Instrument : Bruker IFS-85   Optical resolution : 2 cm-1
Scans : 32   Sampling technique : KBr pellet

Band maxima with relative intensity :

417	21	431	13	502	8	518	14	540	34
624	35	692	22	731	20	779	92	792	64
801	57	855	16	935	42	1018	20	1075	14
1218	78	1250	57	1268	46	1328	40	1409	62
1511	34	1598	41	1693	100	3058	61		

COMPOUND : **Neburon**

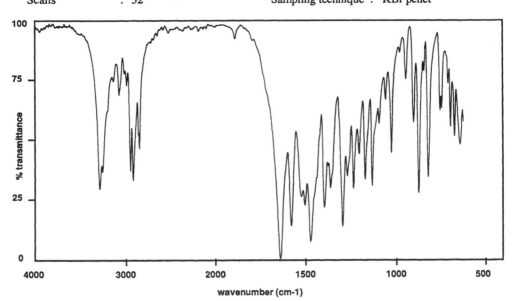

Chemical name   : 1-butyl-3-(3,4-dichlorophenyl)-1-methylurea

Other names   : Kloben, Neburex, Granulex
Type   : herbicide
Brutoformula   : C12H16Cl2N2O   CAS nr   : 555-37-3
Molecular mass : 275.18012   Exact mass   : 274.0639603
Instrument   : Bruker IFS-85   Optical resolution : 2 cm-1
Scans   : 32   Sampling technique : KBr pellet

Band maxima with relative intensity :

643	51	673	47	695	43	709	25	755	36
819	65	874	71	904	42	946	23	1028	54
1059	32	1094	42	1133	68	1172	66	1206	55
1237	70	1271	64	1297	86	1365	69	1399	78
1474	92	1506	77	1581	86	1641	100	1900	6
2861	53	2928	66	2958	62	3003	26	3086	30
3300	70								

COMPOUND : **Nicotine**

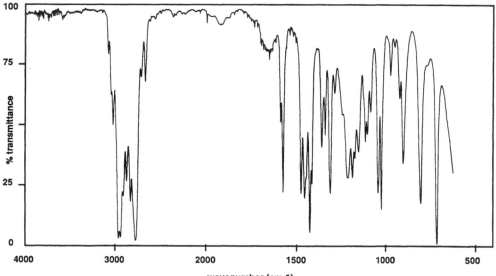

Chemical name : (S)-3-(1-methylpyrrolidin-2-yl) pyridine

Other names : -
Type : insecticide
Brutoformula : C10H14N2
Molecular mass : 162.23648
Instrument : Bruker IFS-85
Scans : 32

CAS nr : 54-11-5
Exact mass : 162.1156908
Optical resolution : 2 cm-1
Sampling technique : neat film

Band maxima with relative intensity :

716	100	807	82	903	65	921	38	971	28
1025	85	1045	78	1086	43	1115	56	1155	60
1190	71	1214	71	1289	36	1314	78	1343	53
1362	58	1427	94	1457	80	1477	78	1576	77
1590	48	1652	21	1918	7	2669	31	2778	98
2833	81	2876	73	2968	97	3029	49		

COMPOUND : **Nitrofen**

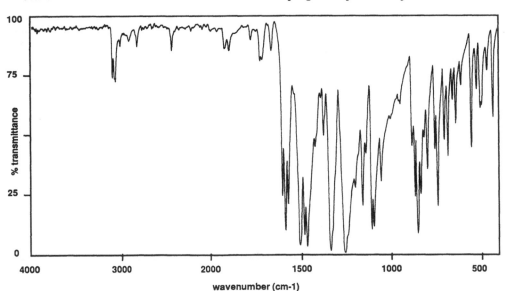

| Chemical name | : 2,4-dichlorophenyl-4-nitrophenyl ether |
| | |

Other names	: TOK E-25, Trizilin		
Type	: herbicide		
Brutoformula	: C12H7Cl2NO3	CAS nr	: 1836-75-5
Molecular mass	: 284.10049	Exact mass	: 282.9802939
Instrument	: Bruker IFS-85	Optical resolution	: 2 cm-1
Scans	: 32	Sampling technique	: KBr pellet

Band maxima with relative intensity :

438	42	470	22	508	38	528	30	557	55
614	29	642	45	659	35	685	59	706	52
743	80	757	56	798	64	835	75	853	91
867	76	885	54	1057	69	1098	88	1110	90
1163	79	1262	100	1342	99	1381	50	1472	97
1488	92	1512	96	1577	79	1594	90	1611	75
1671	13	1733	18	1901	13	2444	13	2833	11
3078	26	3111	24						

COMPOUND : **Nitrothal-isopropyl**

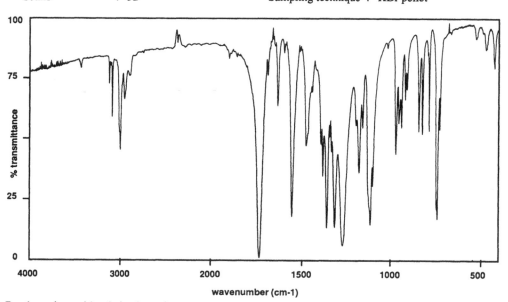

Chemical name   : diisopropyl-5-nitroisophthalate

Other names   : Nitrothal, Pallinal, BAS 3000F
Type   : fungicide
Brutoformula   : C14H17NO6   CAS nr   : 10552-74-6
Molecular mass   : 295.29469   Exact mass   : 295.105576
Instrument   : Bruker IFS-85   Optical resolution : 2 cm-1
Scans   : 32   Sampling technique : KBr pellet

Band maxima with relative intensity :

463	11	713	45	726	83	774	46	811	47
831	46	895	28	905	32	925	44	940	42
957	55	1085	69	1099	85	1144	44	1164	63
1258	94	1302	86	1349	87	1372	65	1383	51
1465	52	1542	82	1623	35	1677	22	1721	100
2936	33	2980	54	3079	40	3112	26		

COMPOUND : **Omethoate**

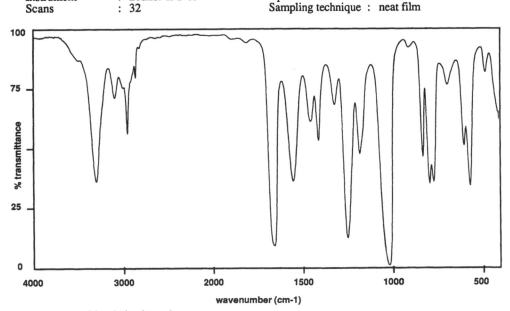

Chemical name	: O,O-dimethyl-S-methylcarbamoylmethyl phosphorothioate		

Other names	: Folimat, S 6876, E 45432		
Type	: acaricide, insecticide		
Brutoformula	: C5H12NO4PS	CAS nr	: 1113-02-6
Molecular mass	: 213.18945	Exact mass	: 213.0224615
Instrument	: Bruker IFS-85	Optical resolution	: 4 cm-1
Scans	: 32	Sampling technique	: neat film

Band maxima with relative intensity :

480	18	572	66	603	49	690	23	773	64
796	65	831	54	1024	100	1182	52	1255	87
1319	31	1411	46	1454	38	1556	63	1664	90
2954	42	3089	26	3305	62				

COMPOUND : **Orbencarb**

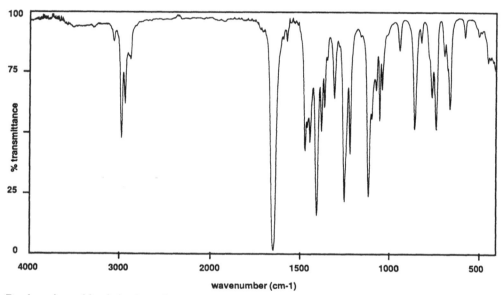

Chemical name	:	2-chlorobenzyl-N,N-diethyl-thiocarbamate

Other names	:	Orthobencarb			
Type	:	herbicide			
Brutoformula	:	C12H16ClSNO	CAS nr	:	-
Molecular mass	:	257.78042	Exact mass	:	257.0641067
Instrument	:	Bruker IFS-85	Optical resolution	:	2 cm-1
Scans	:	32	Sampling technique	:	neat film

Band maxima with relative intensity :

580	10	663	40	740	49	763	35	823	12
860	48	943	15	1037	32	1053	45	1116	77
1218	59	1251	79	1307	35	1363	39	1380	49
1407	85	1444	54	1471	57	1650	100	2935	37
2976	52								

COMPOUND : **Oxadixyl**

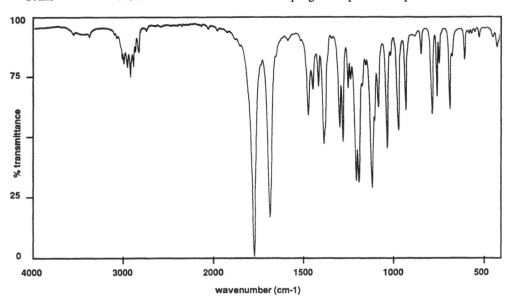

Chemical name	:	2-methoxy-N-(2-oxo-1,3-oxazolidin-3-yl) acet-2,6'-xylidide						

Other names : SAN 371F, Sandofan
Type : fungicide
Brutoformula : C14H18N2O4   CAS nr : 77732-09-3
Molecular mass : 278.31056   Exact mass : 278.1266456
Instrument : Bruker IFS-85   Optical resolution : 2 cm-1
Scans : 32   Sampling technique : KBr pellet

Band maxima with relative intensity :

423	12	603	17	689	38	744	19	757	33
786	40	845	15	932	38	973	47	1034	54
1082	37	1123	71	1194	69	1210	68	1235	25
1249	29	1281	51	1299	45	1388	52	1416	28
1447	29	1475	40	1699	83	1788	100	2820	12
2882	18	2920	23	2952	18				

COMPOUND : **Oxamyl**

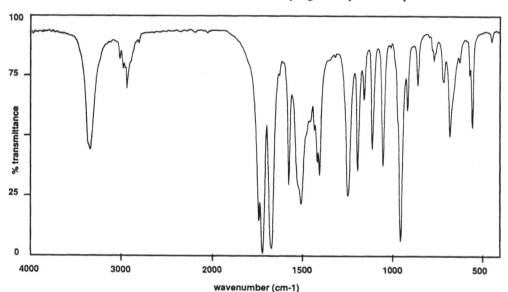

Chemical name : N,N-dimethyl-2-methylcarbamoyloxyimino-2-(methylthio) acetamide

Other names : Vydate, Thioxamyl, DPX 1410
Type : acaricide, insecticide, nematicide
Brutoformula : C7H13N3O3S   CAS nr   : 23135-22-0
Molecular mass : 219.25996   Exact mass   : 219.0677544
Instrument : Bruker IFS-85   Optical resolution : 2 cm-1
Scans : 32   Sampling technique : KBr pellet

Band maxima with relative intensity :

551	46	674	50	708	27	764	18	854	28
910	39	947	94	1048	62	1110	55	1157	34
1191	64	1245	75	1401	66	1501	79	1569	70
1664	97	1715	100	1734	86	2933	29	3329	55

COMPOUND : **Oxycarboxin**

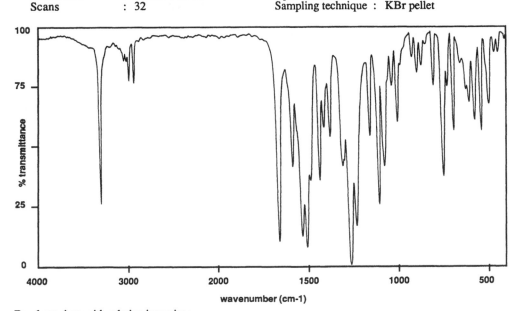

Chemical name	: 2,3-dihydro-6-methyl-5-phenylcarbamoyl-1,4-oxathiin-4,4-dioxide		

Other names	: Plantvax, F 461		
Type	: fungicide		
Brutoformula	: C12H13NO4S	CAS nr	: 5259-88-1
Molecular mass	: 267.30171	Exact mass	: 267.0565221
Instrument	: Bruker IFS-85	Optical resolution	: 2 cm-1
Scans	: 32	Sampling technique	: KBr pellet

Band maxima with relative intensity :

551	46	674	50	708	27	764	18	854	28
910	39	947	94	1048	62	1110	55	1157	34
1191	64	1245	75	1401	66	1501	79	1569	70
1664	97	1715	100	1734	86	2933	29	3329	55

COMPOUND : **Oxydemeton-methyl**

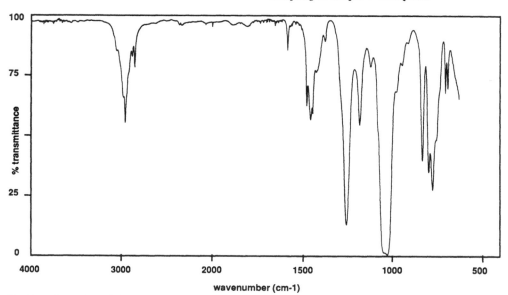

Chemical name : S-2-ethylsulphinylethyl O,O-dimethyl phosphorothioate

Other names : Metasystox R
Type : acaricide, insecticide
Brutoformula : C6H15O4PS2   CAS nr : 301-12-2
Molecular mass : 246.27781   Exact mass : 246.0149384
Instrument : Bruker IFS-85   Optical resolution : 2 cm-1
Scans : 32   Sampling technique : KBr pellet

Band maxima with relative intensity :

687	29	701	31	772	72	793	65	830	60
1023	100	1183	45	1257	87	1457	42	1478	36
1583	13	2849	20	2952	43				

298

COMPOUND : **Oxyfluorfen**

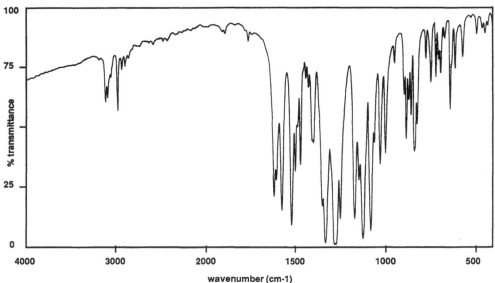

Chemical name : 2-chloro-1-(3-ethoxy-4-nitrophenoxy)-4-(trifluormethyl)benzene

Other names : Goal, Koltar, RH 2915
Type : herbicide
Brutoformula : C15H11ClF3NO4   CAS nr : 42874-03-3
Molecular mass : 361.70742   Exact mass : 361.0328599
Instrument : Bruker IFS-85   Optical resolution : 2 cm-1
Scans : 32   Sampling technique : KBr pellet

Band maxima with relative intensity :

447	10	494	10	572	20	614	24	643	42
695	26	722	28	749	30	777	20	825	48
840	60	860	44	874	38	887	55	898	36
951	22	999	61	1028	65	1081	94	1124	97
1146	72	1169	88	1247	88	1279	100	1330	99
1399	56	1429	33	1442	29	1472	65	1501	68
1522	91	1578	85	1623	79	2985	42	3120	39

COMPOUND : **Paraoxon-ethyl**

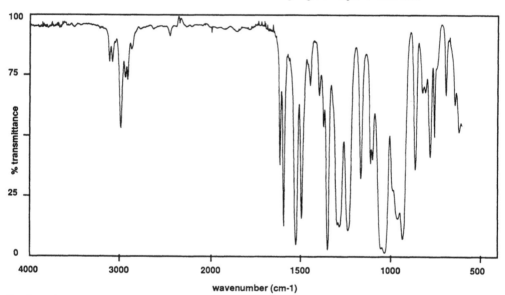

Chemical name : O,O-diethyl O-4-nitrophenyl phosphorate

Other names	: -		
Type	: metabolite of Parathion-ethyl		
Brutoformula	: C10H14NO6P	CAS nr	: -
Molecular mass	: 275.1999	Exact mass	: 275.055866
Instrument	: Bruker IFS-85	Optical resolution	: 2 cm-1
Scans	: 32	Sampling technique	: neat film

Band maxima with relative intensity :

639	37	688	33	752	51	776	59	819	32
860	64	931	94	1032	100	1110	62	1164	68
1236	90	1283	88	1349	98	1370	46	1394	33
1444	29	1492	85	1524	96	1593	88	1613	62
2912	26	2986	47	3082	19	3114	19		

COMPOUND : **Paraquat (dichloride)**

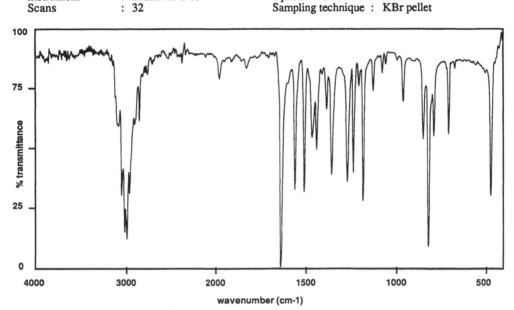

Chemical name   : 1,1'-dimethyl-4,4'-bipyridylium dichloride

Other names   : Paraquat, Gramoxone, R 9910, Dextrone X, Pillarxone
Type   : herbicide
Brutoformula   : C12H14Cl2N2   CAS nr   : 1910-42-5
Molecular mass   : 257.16478   Exact mass   : 256.053397
Instrument   : Bruker IFS-85   Optical resolution : 2 cm-1
Scans   : 32   Sampling technique : KBr pellet

Band maxima with relative intensity :

469	69	706	43	785	44	814	91	845	45
958	29	1074	17	1125	24	1180	71	1205	23
1235	5	1269	63	1355	60	1384	32	1438	49
1464	44	1508	67	1562	67	1640	100	1834	15
1979	19	2857	37	2966	68	2994	87	3019	85
3054	69	3096	40						

COMPOUND : **Parathion**

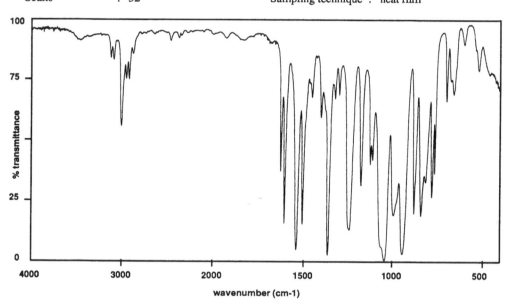

| Chemical name | : O,O-diethyl O-4-nitrophenyl phosphorothioate |
</p>

Other names	: Folidol, Niran, Fosferno
Type	: insecticide, acaricide
Brutoformula	: C10H14NO5PS
Molecular mass	: 291.26054
Instrument	: Bruker IFS-85
Scans	: 32

CAS nr	: 56-38-2
Exact mass	: 291.0330248
Optical resolution	: 2 cm-1
Sampling technique	: neat film

Band maxima with relative intensity :

513	19	595	8	649	29	688	32	749	63
764	73	823	81	861	80	923	97	973	81
1024	100	1098	57	1110	59	1162	68	1225	87
1291	30	1314	31	1348	98	1391	39	1442	31
1490	85	1523	95	1591	84	1614	62	2906	24
2937	24	2984	44	3081	16				

COMPOUND : **Parathion-methyl**

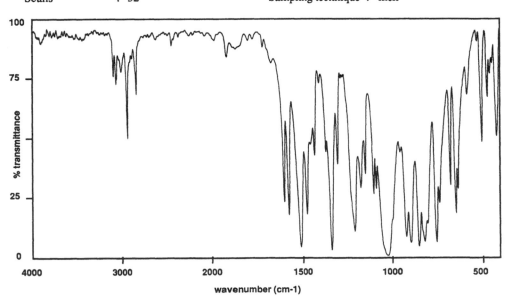

| Chemical name | : O,O-dimethyl O-4-nitrophenyl phosphorothioate |

COMPOUND : **Parathion-methyl**

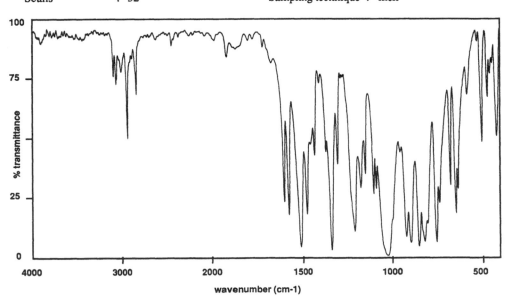

Chemical name : O,O-dimethyl O-4-nitrophenyl phosphorothioate

Other names : Metaphos, Folidol-M, Metacide
Type : insecticide, acaricide
Brutoformula : C8H10NO5PS   CAS nr : 298-00-0
Molecular mass : 263.20636   Exact mass : 263.0017264
Instrument : Bruker IFS-85   Optical resolution : 2 cm-1
Scans : 32   Sampling technique : melt

Band maxima with relative intensity :

428	49	475	33	509	52	588	32	645	72
657	82	686	70	748	78	766	94	832	94
864	96	908	94	933	92	1035	100	1096	71
1109	73	1155	65	1182	71	1218	89	1311	60
1349	97	1440	56	1488	81	1525	95	1590	82
1615	76	1921	14	2848	29	2956	48	3018	20
3079	25	3107	21						

COMPOUND : **Penconazol**

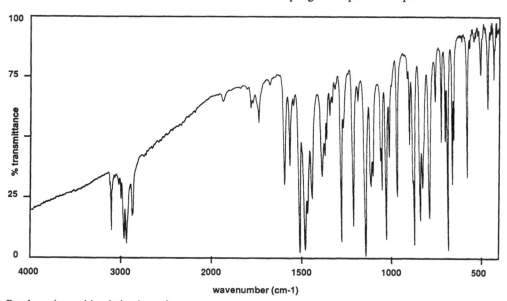

Chemical name   : 1-[2-(2,4-dichlorophenyl)-n-pentyl]-1,2,4-triazole

Other names   : Topaz, CGA 71818
Type   : herbicide
Brutoformula   : C13H15Cl2N3   CAS nr   : -
Molecular mass : 284.1906   Exact mass   : 283.0642948
Instrument   : Bruker IFS-85   Optical resolution : 2 cm-1
Scans   : 32   Sampling technique : KBr pellet

Band maxima with relative intensity :

422	25	456	38	572	67	647	51	654	69
680	97	692	51	714	52	747	35	783	84
820	71	834	85	867	95	891	52	959	74
1003	58	1023	92	1046	71	1054	59	1099	65
1110	70	1143	100	1182	34	1211	87	1278	93
1338	40	1357	50	1383	65	1438	75	1479	97
1508	98	1559	61	1590	69	1730	43	1772	36
2872	81	2935	93	2962	91	2993	74	3103	87

COMPOUND : **Pencycuron**

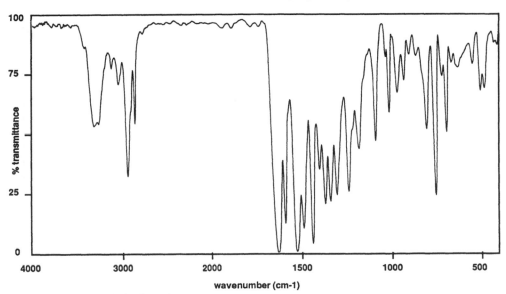

Chemical name   : 1-(4-chlorobenzyl)-1-cyclopentyl-3-phenylurea

Other names   : Monceren
Type   : fungicide
Brutoformula   : C19H21ClN2O
Molecular mass   : 328.84502
Instrument   : Bruker IFS-85
Scans   : 32

CAS nr   : 66063-05-6
Exact mass   : 328.1342302
Optical resolution : 2 cm-1
Sampling technique : KBr pellet

Band maxima with relative intensity :

503	31	545	20	628	21	694	49	756	76
804	47	900	15	931	26	968	32	1012	40
1091	52	1184	56	1244	74	1309	75	1346	78
1375	79	1404	64	1444	96	1494	89	1533	99
1596	87	1637	100	2869	44	2956	66	3053	27
3323	45								

305

COMPOUND : **Pentachlorothioanisol**

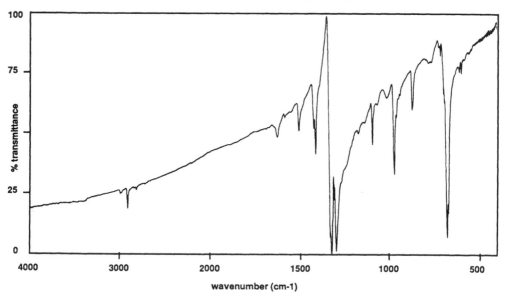

Chemical name   : 2,3,4,5,6-pentachloro thioanisol

Other names   : -
Type   : metabolite of Quintozene
Brutoformula   : C7H3Cl5S
Molecular mass   : 296.4269   CAS nr   : -
Instrument   : Bruker IFS-85   Exact mass   : 293.83981
Scans   : 32   Optical resolution : 2 cm-1
                                      Sampling technique : KBr pellet

Band maxima with relative intensity :

686	93	718	17	875	39	973	66	1097	54
1306	99	1316	77	1333	100	1419	57	1513	48
1634	50	2918	80						

306

COMPOUND : **Pentachlorophenol**

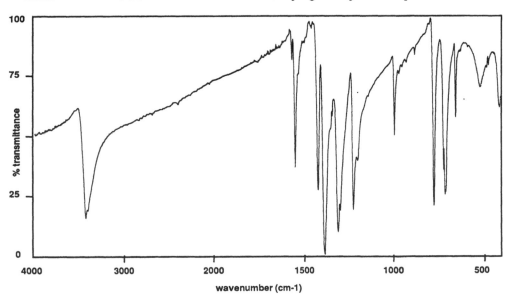

Chemical name	: 1,2,3,4,5-pentachlorophenol		
Other names	: PCP		
Type	: fungicide		
Brutoformula	: C6H5ClO	CAS nr	: 87-86-5
Molecular mass	: 266.33927	Exact mass	: 263.8470042
Instrument	: Bruker IFS-85	Optical resolution	: 2 cm-1
Scans	: 32	Sampling technique	: KBr pellet

Band maxima with relative intensity :

514	29	649	42	710	74	717	64	770	79
986	49	1219	81	1305	90	1380	100	1418	72
1546	62	1563	17	3414	83				

307

COMPOUND : **Permethrin**

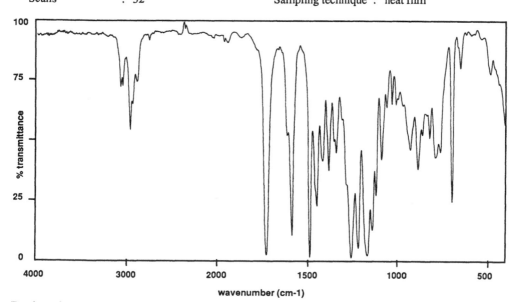

Chemical name	:	3-phenoxybenzyl (1RS)-cis,trans-3-(2,2-dichlorovinyl-2,2-dimethylcyclopropanecarboxylate
Other names	:	Ambush, Ectiban, Perthrin, RP 557
Type	:	insecticide, repellent

Brutoformula	:	C21H20Cl2O3	CAS nr	:	52645-53-1
Molecular mass	:	391.29775	Exact mass	:	390.0789405
Instrument	:	Bruker IFS-85	Optical resolution	:	2 cm-1
Scans	:	32	Sampling technique	:	neat film

Band maxima with relative intensity :

483	22	647	19	692	76	784	57	816	49
880	62	922	54	999	35	1023	34	1054	36
1084	58	1115	73	1137	88	1166	99	1217	95
1257	100	1340	55	1380	62	1414	58	1447	77
1488	99	1585	90	1727	98	2955	45	3064	27

308

COMPOUND : **Perthane**

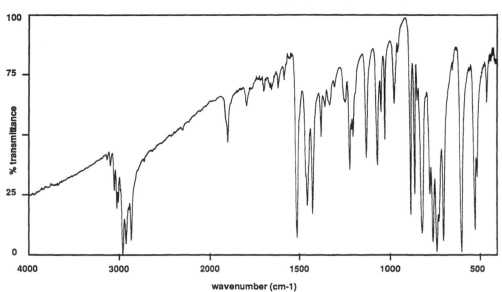

Chemical name	: 1,1-dichloro-2,2-bis (4-ethylphenyl)-ethane		
Other names	: Ethylan, Q 137		
Type	: insecticide		
Brutoformula	: C18H20Cl2	CAS nr	: 72-56-0
Molecular mass	: 307.2661	Exact mass	: 306.0941982
Instrument	: Bruker IFS-85	Optical resolution	: 2 cm-1
Scans	: 32	Sampling technique	: KBr pellet

Band maxima with relative intensity :

452	36	510	67	523	90	598	100	698	95
735	99	756	95	771	75	816	92	857	75
881	83	968	36	1019	51	1040	40	1063	62
1124	59	1195	50	1216	64	1238	35	1323	36
1371	50	1422	83	1451	79	1511	93	1578	25
1612	29	1650	30	1692	30	1789	36	1896	52
2867	93	2925	95	2960	99	3024	80	3051	72
3091	61								

309

COMPOUND : **Phenmedipham**

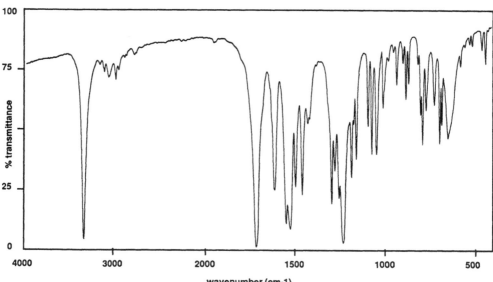

Chemical name   : methyl-3-m-tolylcarbamoyloxyphenylcarbamate

Other names   : Betanal, SN 38584
Type   : herbicide
Brutoformula   : C16H16N2O4   CAS nr   : 13684-63-4
Molecular mass   : 300.31692   Exact mass   : 300.1109964
Instrument   : Bruker IFS-85   Optical resolution : 2 cm-1
Scans   : 32   Sampling technique : KBr pellet

Band maxima with relative intensity :

439	22	459	17	647	54	681	48	692	56
722	40	768	42	787	56	798	44	864	31
879	37	894	22	928	31	1002	41	1039	60
1067	60	1088	49	1152	62	1178	70	1225	98
1247	79	1270	67	1288	81	1420	47	1452	77
1489	74	1519	92	1542	90	1607	75	1712	100
2953	28	3035	27	3323	96					

COMPOUND : **Phenothrin**

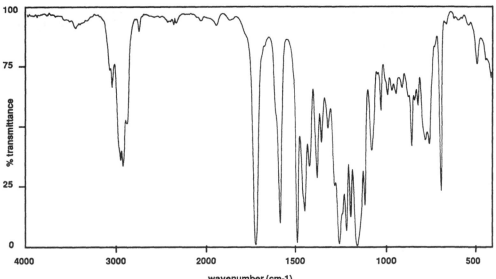

Chemical name   : 3-phenoxybenzyl-d-cis,trans chrysanthemate

Other names   : Sumithrin, S 2539
Type   : insecticide
Brutoformula   : C20H24O3   CAS nr   : 26002-80-2
Molecular mass   : 312.41248   Exact mass   : 312.1725327
Instrument   : Bruker IFS-85   Optical resolution : 2 cm-1
Scans   : 32   Sampling technique : neat film

Band maxima with relative intensity :

486	22	692	76	756	56	817	40	855	57
940	35	986	35	1023	42	1076	59	1115	82
1158	100	1192	87	1216	93	1257	99	1318	50
1353	56	1379	71	1420	66	1447	85	1488	98
1585	90	1726	99	2735	8	2924	65	3038	31

COMPOUND : **Phenthoate**

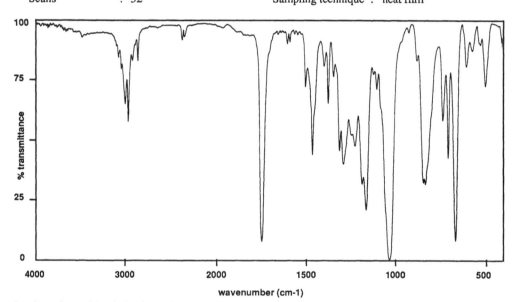

Chemical name : S-α-ethoxycarbonylbenzyl O,O-dimethylphosphorodithioate

Other names : Fenthoate, Cidial, Elsan
Type : acaricide, insecticide
Brutoformula : C12H17O4PS2   CAS nr : 2597-03-7
Molecular mass : 320.36065   Exact mass : 320.030584
Instrument : Bruker IFS-85   Optical resolution : 2 cm-1
Scans : 32   Sampling technique : neat film

Band maxima with relative intensity :

494	26	599	18	655	92	697	57	728	41
822	68	1018	100	1094	27	1152	79	1215	51
1280	59	1302	54	1338	22	1367	33	1391	18
1454	55	1495	27	1599	8	1737	92	2843	16
2948	41	2983	34						

COMPOUND : **O-Phenylphenol**

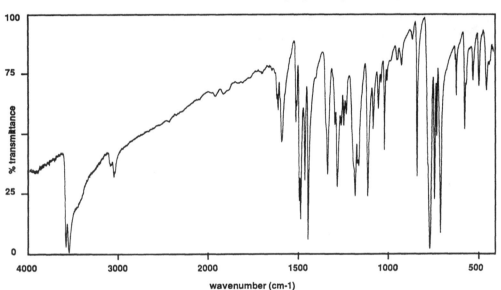

Chemical name : 2-phenylphenol

Other names : Brunosol, Dowicide
Type : fungicide
Brutoformula : C12H10O
Molecular mass : 170.2129
Instrument : Bruker IFS-85
Scans : 32

CAS nr : 90-43-7
Exact mass : 170.0731601
Optical resolution : 2 cm-1
Sampling technique : KBr pellet

Band maxima with relative intensity :

445	30	487	28	521	26	566	47	614	32
699	91	721	50	732	77	757	98	828	67
915	20	1008	56	1043	38	1071	47	1100	75
1150	62	1170	75	1218	40	1232	45	1269	72
1283	45	1324	66	1433	94	1453	69	1476	85
1484	77	1505	44	1585	52	1606	39	3028	67
3525	100	3563	97						

COMPOUND : **Phorate**

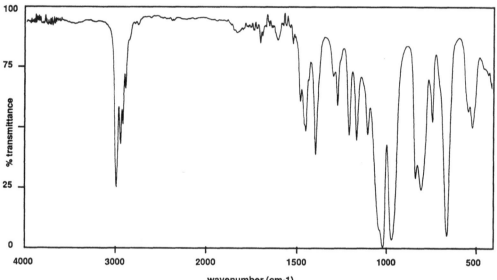

Chemical name	: O,O-diethyl-S-ethylthiomethyl phosphorodithioate		

Other names	: Thimet, Rampart, Agrimet		
Type	: acaricide, insecticide		
Brutoformula	: C7H17O2PS3	CAS nr	: 298-02-2
Molecular mass	: 260.3661	Exact mass	: 260.0128285
Instrument	: Bruker IFS-85	Optical resolution	: 2 cm-1
Scans	: 32	Sampling technique	: neat film

Band maxima with relative intensity :

402	37	514	49	655	95	731	47	797	75
827	70	961	96	1013	100	1098	52	1160	54
1202	52	1267	39	1388	60	1442	50	1473	38
1596	12	1690	13	2870	32	2901	47	2928	55
2978	74								

COMPOUND : **Phosalone**

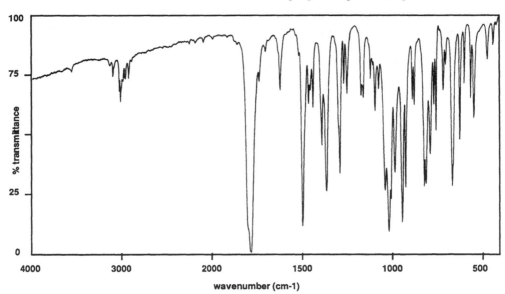

Chemical name	: S-6-chloro-2,3-dihydro-2-oxobenzoxazol-3-ylmethyl-O,O-diethylphosphorodithioate		
Other names	: Zolone, Rubitox, Azofene		
Type	: acaricide, insecticide		
Brutoformula	: C12H15ClNO4PS2	CAS nr	: 2310-17-0
Molecular mass	: 367.80441	Exact mass	: 366.9868611
Instrument	: Bruker IFS-85	Optical resolution	: 2 cm-1
Scans	: 32	Sampling technique	: KBr pellet

Band maxima with relative intensity :

430	11	462	18	536	43	553	34	590	28
615	52	655	72	705	31	744	48	756	37
775	58	798	70	808	72	865	37	877	35
914	72	933	87	972	66	1004	91	1026	73
1064	30	1083	40	1110	26	1151	34	1242	32
1259	28	1280	66	1353	74	1380	54	1431	38
1454	36	1487	88	1614	31	1774	100	2902	25
2995	35	3076	25						

315

COMPOUND : **Phosmet**

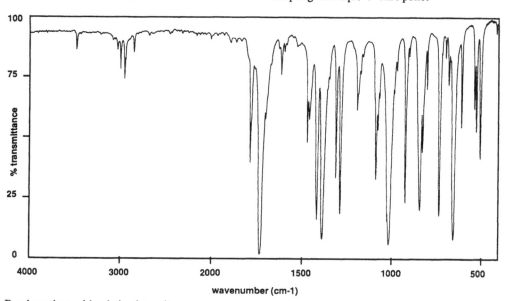

Chemical name : O,O-dimethyl-S-phthalimidomethyl phosphorodithioate

Other names : Imidan, Prolate, Kemolate
Type : insecticide, acaricide
Brutoformula : C11H12NO4PS2   CAS nr : 732-11-6
Molecular mass : 317.31635   Exact mass : 316.9945342
Instrument : Bruker IFS-85   Optical resolution : 2 cm-1
Scans : 32   Sampling technique : KBr pellet

Band maxima with relative intensity :

498	58	520	47	530	37	604	45	648	93
676	27	692	16	726	83	797	29	821	56
835	80	916	77	1004	95	1083	67	1189	38
1280	82	1305	67	1379	93	1407	85	1453	42
1464	52	1612	23	1726	100	1781	61	2848	14
2948	25	2995	21	3482	13				

COMPOUND : **Phosphamidon**

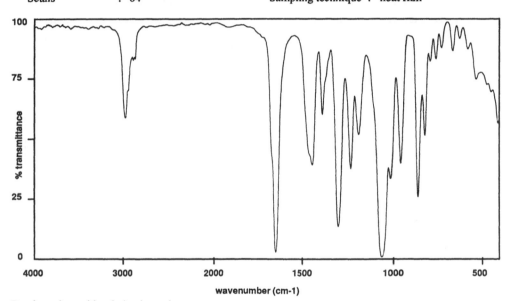

Chemical name : 2-chloro-2-diethylcarbamoyl-1-methylvinyl dimethylphosphate

Other names : Dimecron, Apamidon, Dixon
Type : acaricide, insecticide
Brutoformula : C10H19ClNO5P   CAS nr   : 13171-21-6
Molecular mass : 299.69339   Exact mass   : 299.0689282
Instrument : Bruker IFS-85   Optical resolution : 4 cm-1
Scans : 64   Sampling technique : neat film

Band maxima with relative intensity :

659	12	723	10	756	15	817	48	856	74
952	60	1051	100	1180	47	1224	62	1292	87
1382	39	1438	60	1643	98	2970	40		

COMPOUND : **Phoxim**

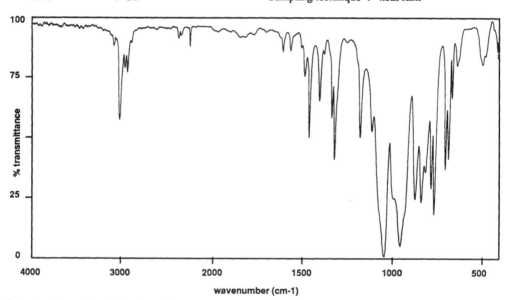

| Chemical name | : O,O-diethyl-α-cyanobenzylideneamino-oxyphosphonothioate |
| | |

Other names : Foxim, Valexon, Baythion, Volaton
Type : insecticide

Brutoformula	: C12H15N2O3PS	CAS nr	: 14816-18-3
Molecular mass	: 298.29871	Exact mass	: 298.0540944
Instrument	: Bruker IFS-85	Optical resolution	: 2 cm-1
Scans	: 32	Sampling technique	: neat film

Band maxima with relative intensity :

487	18	627	19	654	32	671	58	688	62
750	81	768	70	822	76	856	75	934	95
1023	100	1098	47	1163	49	1303	59	1321	41
1390	34	1446	49	1474	24	1555	13	1598	13
2236	12	2907	23	2984	43				

318

COMPOUND : **Piperonyl butoxide**

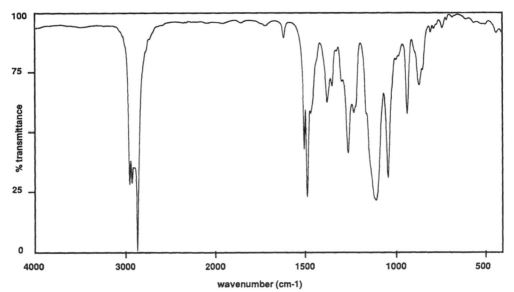

Chemical name : 5-[2-(2-butoxyethoxy) ethoxymethyl]-6-propyl-1,3-benzodioxole

Other names : Butacide, Butoxide
Type : synergist
Brutoformula : C19H30O5   CAS nr : 51-03-6
Molecular mass : 338.44795   Exact mass : 338.2093085
Instrument : Bruker IFS-85   Optical resolution : 2 cm-1
Scans : 32   Sampling technique : neat film

Band maxima with relative intensity :

405	8	866	29	935	42	1039	69	1104	78
1258	58	1377	37	1486	77	1503	57	1620	9
2869	100	2931	71	2957	71				

COMPOUND : **KW-139**

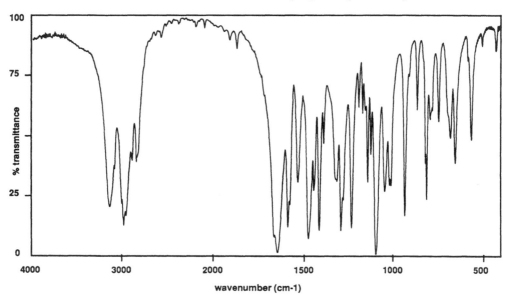

Chemical name	:	ethyl-3-chloro-2,6-dimethoxybenzohydroxamate

Other names	: -				
Type	: metabolite of Benzomate				
Brutoformula	: C11H14ClNO4		CAS nr	:	-
Molecular mass	: 259.6915		Exact mass	:	259.061127
Instrument	: Bruker IFS-85		Optical resolution	:	2 cm-1
Scans	: 32		Sampling technique	:	KBr pellet

Band maxima with relative intensity :

429	13	504	11	564	51	651	61	676	50
741	43	789	42	809	76	816	61	861	38
929	83	1005	70	1042	73	1092	100	1121	57
1139	69	1168	40	1190	37	1230	88	1290	89
1313	69	1388	53	1412	89	1443	72	1473	93
1534	69	1589	88	1646	99	1868	12	1907	9
2094	4	2193	3	2846	61	2891	60	2983	88
3142	80								

COMPOUND : **Pirimicarb**

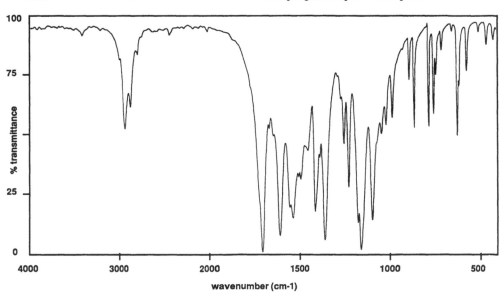

| Chemical name | : 2-dimethylamino-5,6-dimethylpyrimidin-4-yl dimethylcarbamate |
| | |

Other names   : PP 062, Pirimor, Rapid, Aphox
Type   : insecticide
Brutoformula   : C11H18N4O2   CAS nr   : 23103-98-2
Molecular mass   : 238.29171   Exact mass   : 238.1429638
Instrument   : Bruker IFS-85   Optical resolution : 2 cm-1
Scans   : 32   Sampling technique : KBr pellet

Band maxima with relative intensity :

424	9	461	11	572	23	625	50	713	14
743	24	756	41	782	46	863	47	890	26
981	42	1016	45	1092	86	1154	98	1223	72
1249	54	1354	94	1408	82	1536	85	1612	93
1709	100	2867	37	2930	47				

COMPOUND : **Pirimiphos-ethyl**

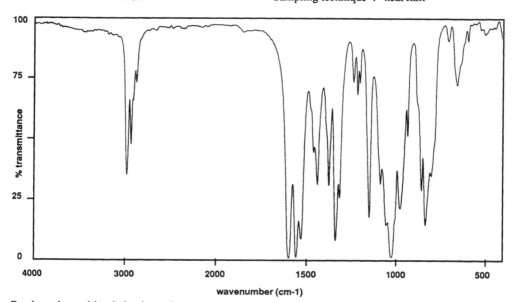

Chemical name   : O-2-diethylamino-6-methylpyrimidin-4-yl O,O-diethylphosphorothioate

Other names	: Primicid, PP 211			
Type	: acaricide, insecticide			
Brutoformula	: C13H24N3O3PS	CAS nr	:	23505-41-1
Molecular mass	: 333.38829	Exact mass	:	333.127589
Instrument	: Bruker IFS-85	Optical resolution	:	2 cm-1
Scans	: 32	Sampling technique	:	neat film

Band maxima with relative intensity :

405	7	501	5	654	26	702	8	832	86
853	71	931	48	972	79	1024	99	1086	68
1150	82	1216	30	1239	25	1317	74	1339	92
1375	69	1438	68	1528	91	1555	99	1596	100
2932	51	2978	64						

COMPOUND : **Pirimiphos-methyl**

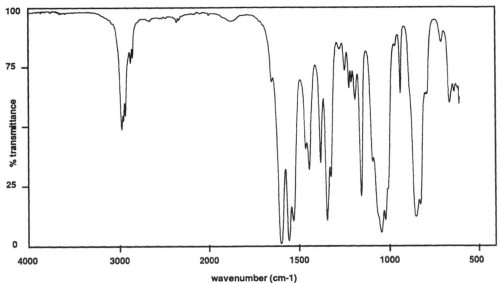

Chemical name : O-2-diethylamino-6-methylpyrimidin-4-yl O,O-dimethylphosphorothioate

Other names : PP 511, Actellic, Blex, Silosan
Type : acaricide, insecticide
Brutoformula : C11H20N3O3PS
Molecular mass : 305.33411
Instrument : Bruker IFS-85
Scans : 32

CAS nr : 23505-41-1
Exact mass : 305.0962906
Optical resolution : 2 cm-1
Sampling technique : neat film

Band maxima with relative intensity :

655	40	847	88	931	36	1041	95	1150	80
1182	38	1216	33	1239	26	1339	90	1375	65
1438	68	1529	90	1555	98	1598	100	2932	44
2973	50								

COMPOUND : **Plifenate**

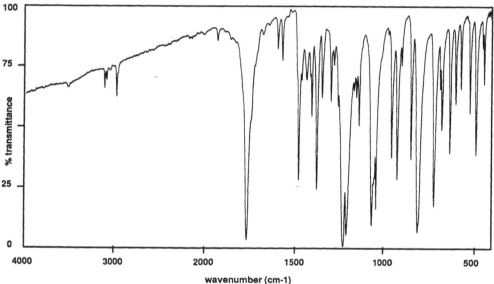

Chemical name   : 2,2,2,-trichloro-1-(3,4-dichlorophenyl) ethyl acetate

Other names   : Penfenate, Baygon MEB
Type   : insecticide
Brutoformula   : C10H7Cl5O2   CAS nr   : 21757-82-4
Molecular mass : 336.43109   Exact mass   : 333.8888659
Instrument   : Bruker IFS-85   Optical resolution : 2 cm-1
Scans   : 32   Sampling technique : KBr pellet

Band maxima with relative intensity :

440	31	485	61	518	43	566	33	595	39
625	60	670	50	679	33	711	82	801	93
837	62	889	23	912	71	943	62	1029	84
1056	91	1129	48	1143	36	1199	95	1217	100
1271	23	1289	38	1339	37	1369	75	1397	44
1426	29	1470	71	1564	21	1590	16	1759	97
2951	36	3091	33						

COMPOUND : **PP-333**

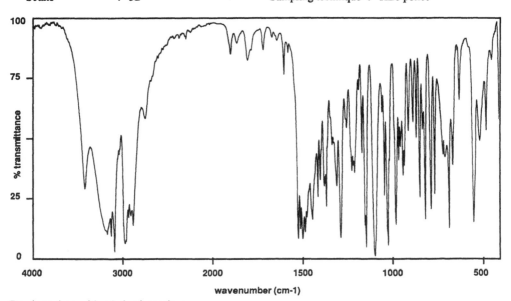

Chemical name   : 1-(4-chlorophenyl)-4,4-dimethyl-2-(1,2,4-triazol-1yl) pentan-3-ol

Other names   : Paclobutrazol
Type   : fungicide, herbicide
Brutoformula   : C15H20ClN3O
Molecular mass   : 293.79915
Instrument   : Bruker IFS-85
Scans   : 32

CAS nr   : -
Exact mass   : 293.1294788
Optical resolution : 2 cm-1
Sampling technique : KBr pellet

Band maxima with relative intensity :

475	46	511	51	546	85	659	61	680	88
713	57	757	73	779	80	810	84	824	46
840	75	860	49	904	47	934	65	947	53
957	59	974	86	1016	95	1035	74	1089	100
1138	96	1162	56	1202	64	1217	63	1281	92
1302	70	1361	78	1396	67	1409	74	1439	84
1480	89	1495	92	1507	88	1521	92	2871	86
2966	94	3083	97	3121	90	3165	90	3412	70

COMPOUND : **Prochloraz**

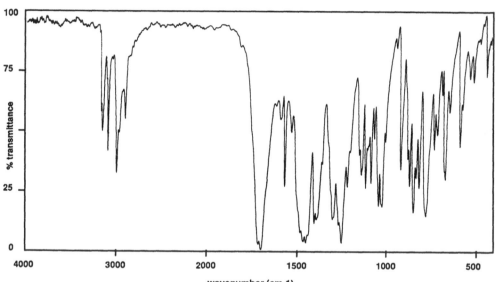

Chemical name	:	N-propyl-N-[2-(2,4,6-trichlorophenoxy) ethyl]-1H-imidazole-1-carboxamide		
Other names	:	Sportak, Sporgon, BTS 40542		
Type	:	fungicide		
Brutoformula	:	C15H16Cl3N3O2	CAS nr	: 67747-09-5
Molecular mass	:	376.67267	Exact mass	: 375.0308007
Instrument	:	Bruker IFS-85	Optical resolution	: 2 cm-1
Scans	:	32	Sampling technique	: KBr pellet

Band maxima with relative intensity :

430	26	501	28	521	26	575	55	633	41
654	69	674	33	699	50	715	57	760	85
799	73	830	83	851	72	901	65	1004	81
1025	81	1052	52	1070	71	1102	73	1125	67
1205	73	1235	96	1286	86	1386	88	1433	96
1515	50	1552	73	1577	44	1681	100	2873	45
2964	67	3063	58	3124	50				

COMPOUND : **Proclonol**

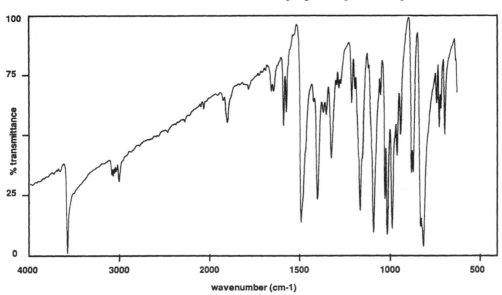

Chemical name   : di-(4-chlorophenyl) cycopropyl methanol

Other names   : -
Type   : insecticide
Brutoformula   : C15H14O2Cl2   CAS nr   : -
Molecular mass   : 297.18363   Exact mass   : 296.0370788
Instrument   : Bruker IFS-85   Optical resolution : 2 cm-1
Scans   : 32   Sampling technique : KBr pellet

Band maxima with relative intensity :

694	49	725	46	739	36	812	97	870	65
879	65	940	49	958	58	985	89	1011	92
1026	77	1051	32	1089	91	1163	82	1210	36
1282	29	1323	59	1352	41	1399	77	1488	87
1573	40	1590	45	1657	31	1902	44	3005	69
3571	100								

327

COMPOUND : **Procymidone**

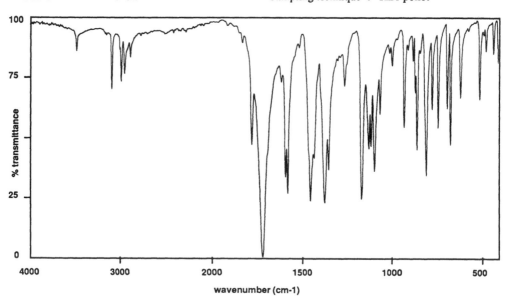

Chemical name   : N-(3,5-dichlorophenyl)-1,2-dimethylcyclopropane-1,2-dicarboximide

Other names   : Sumisclex, Dicyclidine, Sumilex, S 7131
Type   : fungicide
Brutoformula   : C13H11Cl2NO2   CAS nr   : 32809-16-8
Molecular mass   : 284.14412   Exact mass   : 283.0166782
Instrument   : Bruker IFS-85   Optical resolution : 2 cm-1
Scans   : 32   Sampling technique : KBr pellet

Band maxima with relative intensity :

429	14	472	13	506	33	612	32	668	52
686	37	736	45	770	37	802	65	853	54
864	30	875	17	923	45	988	19	1056	39
1086	63	1106	53	1118	54	1159	75	1254	27
1344	63	1367	77	1446	76	1573	73	1586	66
1714	100	1775	52	2871	15	2936	22	2975	25
3084	28								

COMPOUND : **Profenofos**

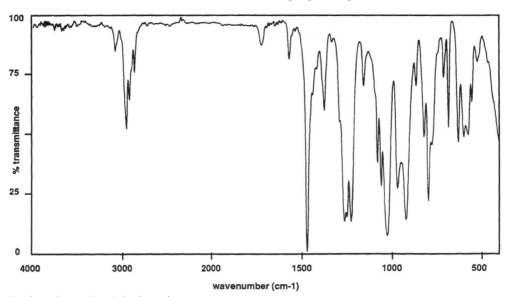

Chemical name   : O-(4-bromo-chloro-) phenyl-O-ethyl-S-propylphosphorothioate

Other names   : Selectron, CGA 15324
Type   : insecticide
Brutoformula   : C11H15BrClPSO3   CAS nr   : -
Molecular mass   : 373.63116   Exact mass   : 371.9351911
Instrument   : Bruker IFS-85   Optical resolution : 2 cm-1
Scans   : 32   Sampling technique : neat film

Band maxima with relative intensity :

525	19	554	36	599	51	629	53	684	47
711	25	796	78	821	51	866	29	923	86
969	73	1027	93	1061	72	1082	62	1162	29
1232	87	1268	87	1379	39	1473	100	1575	17
1731	11	2873	23	2933	34	2966	47	3089	14

329

COMPOUND : **Prometryne**

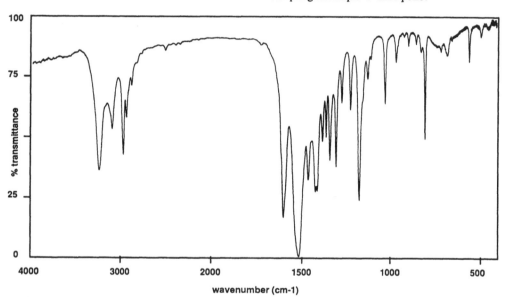

Chemical name   : 2,4-bis(isopropylamino)-6-methylthio-1,3,5-triazine

Other names   : Gesagard, Caparol, Prometrex
Type   : herbicide
Brutoformula   : C10H19N5S   CAS nr   : 7287-19-6
Molecular mass : 241.35643   Exact mass   : 241.1361061
Instrument   : Bruker IFS-85   Optical resolution : 2 cm-1
Scans   : 32   Sampling technique : KBr pellet

Band maxima with relative intensity :

565	18	686	15	809	50	899	11	967	18
1028	35	1128	25	1176	76	1224	38	1274	35
1307	62	1343	59	1364	49	1384	51	1424	72
1464	67	1518	100	1604	83	2930	41	2967	56
3095	45	3242	63						

COMPOUND : **Propachlor**

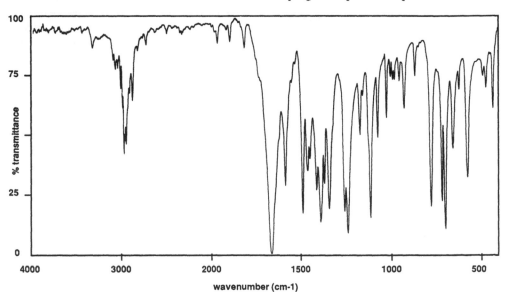

Chemical name	:	2-chloro-N-isopropylacetanilide

Other names	:	Ramrod, CP 31393, Bexton			
Type	:	herbicide			
Brutoformula	:	C11H14ClNO	CAS nr	:	1918-16-7
Molecular mass	:	211.69333	Exact mass	:	211.0763848
Instrument	:	Bruker IFS-85	Optical resolution	:	2 cm-1
Scans	:	32	Sampling technique	:	KBr pellet

Band maxima with relative intensity :

433	39	472	30	577	68	623	30	658	56
703	90	720	78	781	80	868	25	929	38
955	27	992	26	1004	25	1027	42	1075	51
1118	85	1174	50	1244	91	1261	82	1345	81
1370	70	1392	86	1414	73	1463	65	1492	83
1591	71	1670	100	1817	12	1896	9	1964	10
2877	34	2949	52	2974	56	2987	37	3007	29
3061	21								

COMPOUND : **Propamocarb**

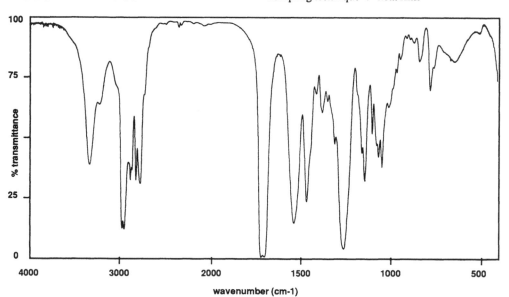

Chemical name   : propyl-3-(dimethylamino) propylcarbamate hydrochloride

Other names   : Prevex, Filex, SN 39744
Type   : fungicide
Brutoformula   : C9H21ClN2O2   CAS nr   : 25606-41-1
Molecular mass   : 224.73292   Exact mass   : 224.1291443
Instrument   : Bruker IFS-85   Optical resolution : 2 cm-1
Scans   : 32   Sampling technique : neat film

Band maxima with relative intensity :

643	17	779	29	839	17	1042	61	1061	57
1100	47	1139	67	1256	96	1378	38	1463	76
1536	85	1721	100	2767	68	2816	67	2878	67
2945	88	3335	60						

332

COMPOUND : **Propargite**

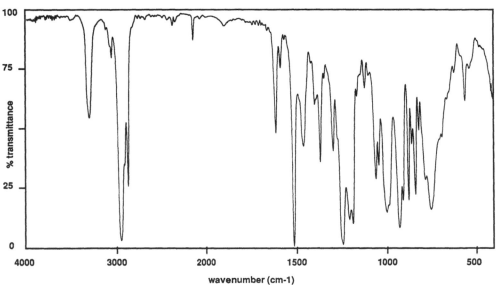

Chemical name   : 2-(4-tert.butylphenoxy) cyclohexyl prop-2-ynyl sulphite

Other names   : Omite, Comite, DO 14
Type   : acaricide
Brutoformula   : C19H26O4S   CAS nr   : 2312-35-8
Molecular mass   : 350.47667   Exact mass   : 350.1551687
Instrument   : Bruker IFS-85   Optical resolution : 2 cm-1
Scans   : 32   Sampling technique : neat film

Band maxima with relative intensity :

550	38	744	85	811	51	831	78	851	56
870	81	902	80	923	92	993	86	1038	65
1055	71	1115	32	1185	90	1240	99	1291	59
1363	64	1455	57	1511	100	1579	23	1608	51
2129	11	2866	73	2948	96	3038	18	3283	44

COMPOUND : **Propazine**

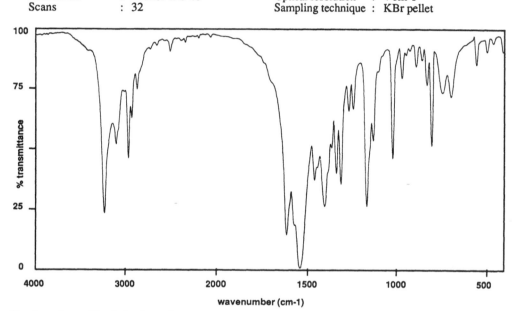

Chemical name	: 2-chloro-4,6-bis(isopropylamino)-1,3,5-triazine		
Other names	: G 30028, Gesamil, Milogard		
Type	: herbicide		
Brutoformula	: C9H16ClN5	CAS nr	: 139-40-2
Molecular mass	: 229.71437	Exact mass	: 229.1094127
Instrument	: Bruker IFS-85	Optical resolution	: 4 cm-1
Scans	: 32	Sampling technique	: KBr pellet

Band maxima with relative intensity :

557	14	698	28	744	26	806	48	831	22
889	15	968	20	1020	5	1130	46	1168	74
1244	32	1269	33	1315	64	1340	60	1407	73
1463	62	1546	100	1622	85	2875	23	2972	53
3112	46	3249	76						

COMPOUND : **Propetamphos**

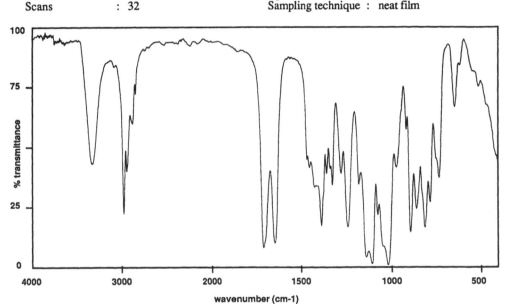

Chemical name : (E)-O-2-isopropoxycarbonyl-1-methylvinyl-O-methyl-ethylphosphoro-
amidothioate
Other names : Propetamphos, Safrotin
Type : insecticide
Brutoformula : C10H20NO4PS   CAS nr   : 31218-83-4
Molecular mass : 281.30896   Exact mass   : 281.0850583
Instrument : Bruker IFS-85   Optical resolution : 2 cm-1
Scans : 32   Sampling technique : neat film

Band maxima with relative intensity :

644	32	731	63	782	73	812	84	860	76
892	86	917	42	972	58	1017	100	1107	99
1180	65	1241	83	1278	61	1325	66	1357	61
1385	83	1648	90	1712	92	2946	59	2979	77
3327	56								

335

COMPOUND : **Propham**

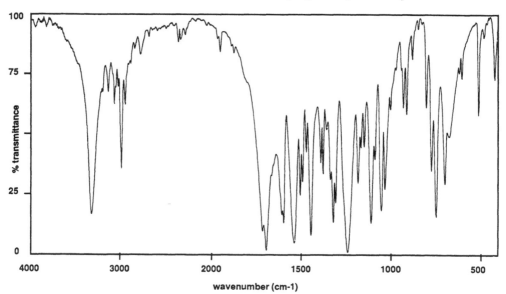

Chemical name	: isopropyl phenylcarbamate		
Other names	: Prophenam, IPC, Tuberite, Premalox		
Type	: herbicide, growth regulator		
Brutoformula	: C10H13NO2	CAS nr	: 122-42-9
Molecular mass	: 179.22061	Exact mass	: 179.0946212
Instrument	: Bruker IFS-85	Optical resolution	: 2 cm-1
Scans	: 32	Sampling technique	: KBr pellet

Band maxima with relative intensity :

418	27	509	42	601	26	695	71	746	85
773	65	800	38	877	18	908	41	925	38
1048	82	1084	60	1107	87	1142	55	1178	70
1238	100	1303	78	1317	87	1373	66	1386	61
1441	92	1467	57	1488	69	1501	75	1537	95
1596	87	1696	98	1949	14	2772	15	2937	36
2980	63	3060	36	3131	31	3317	83		

COMPOUND : **Propiconazole**

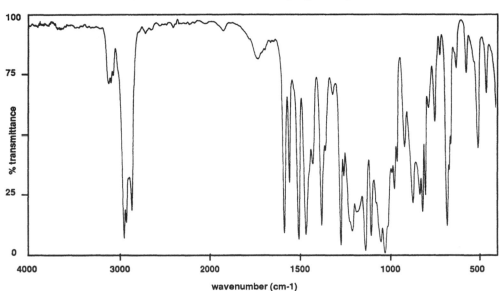

Chemical name : 1-[[2-(2,4-dichlorophenyl)-4-propyl-1,3-dioxolan-2-yl]methyl]-1H-
1,2,4-triazole (cis/trans)
Other names : Proconazol, CGA 64250, Desmel
Type : fungicide
Brutoformula : C15H17Cl2N3O2   CAS nr   : 60207-90-1
Molecular mass : 342.2276   Exact mass   : 341.0697722
Instrument : Bruker IFS-85   Optical resolution : 2 cm-1
Scans : 32   Sampling technique : neat film

Band maxima with relative intensity :

456	32	503	55	565	23	621	21	678	88
710	16	741	44	796	75	813	82	829	75
868	78	913	55	956	61	973	72	1028	100
1050	95	1105	92	1137	98	1210	90	1255	67
1273	96	1378	88	1426	61	1465	92	1506	94
1557	69	1587	91	1727	17	2874	80	2934	86
2959	92	3118	26						

337

COMPOUND : **Propineb**

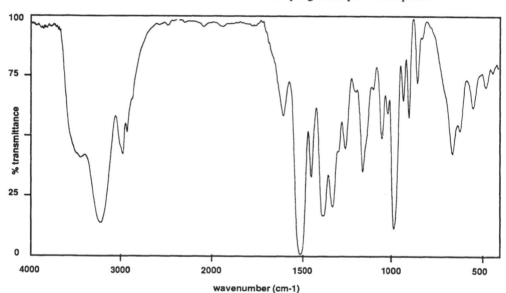

(x> 1)

Chemical name   : zinc propylene bis(dithiocarbamate)

Other names   : Mezineb, Antracol, Airone, Taifen
Type   : acaricide, fungicide
Brutoformula   : (C5H8N2S4ZN)x   CAS nr   : 12071-83-9
Molecular mass   : 289.75291 (monomer)   Exact mass   : 287.886175
Instrument   : Bruker IFS-85   Optical resolution : 2 cm-1
Scans   : 32   Sampling technique : KBr pellet

Band maxima with relative intensity :

481	29	547	38	658	57	852	27	898	42
928	35	979	89	1013	40	1048	50	1157	64
1256	55	1328	79	1386	83	1448	67	1511	100
1606	41	2972	57	3223	86				

338

COMPOUND : **Propoxur**

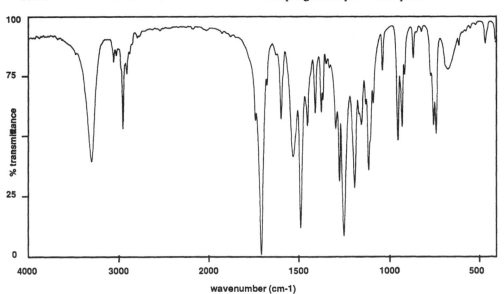

Chemical name	: 2-isopropoxyphenyl methylcarbamate		
Other names	: Arprocarb, Unden, Baygon, OMS 33		
Type	: insecticide		
Brutoformula	: C11H15NO3	CAS nr	: 114-26-1
Molecular mass	: 209.2471	Exact mass	: 209.1051845
Instrument	: Bruker IFS-85	Optical resolution	: 2 cm-1
Scans	: 32	Sampling technique	: KBr pellet

Band maxima with relative intensity :

409	10	467	10	675	21	743	48	757	44
870	16	935	45	957	51	1040	21	1119	64
1159	44	1197	71	1257	92	1281	68	1301	46
1380	39	1414	39	1458	45	1495	88	1538	58
1603	42	1715	100	2974	46	3071	17	3317	59

COMPOUND : **Propylene thiourea**

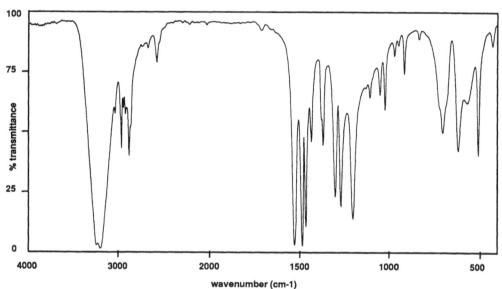

Chemical name   : propylene thiourea

Other names   :
Type   : metabolite of Propineb
Brutoformula   : C4H8SN2   CAS nr   : 2055-46-1
Molecular mass   : 116.18176   Exact mass   : 116.0408159
Instrument   : Bruker IFS-85   Optical resolution : 2 cm-1
Scans   : 32   Sampling technique : KBr pellet

Band maxima with relative intensity :

430	16	512	62	622	60	707	52	915	27
968	19	1020	42	1049	36	1105	37	1204	88
1271	83	1304	79	1371	57	1436	55	1471	91
1491	99	1536	99	2573	21	2887	60	2972	57
3218	100								

COMPOUND : **Propyzamide**

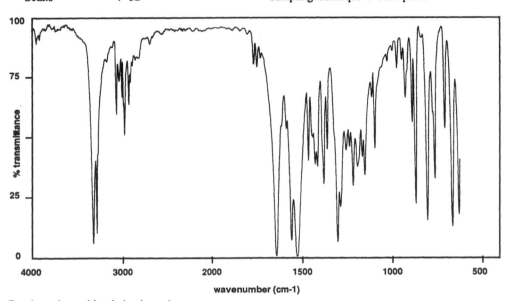

Chemical name : 3,5-dichloro-N-(1,1-dimethyl-2-propynyl) benzamide

Other names : Pronamide, Kerb, Clanex
Type : herbicide
Brutoformula : C12H11Cl2NO   CAS nr : 23950-58-5
Molecular mass : 256.13357   Exact mass : 255.0217641
Instrument : Bruker IFS-85   Optical resolution : 2 cm-1
Scans : 32   Sampling technique : KBr pellet

Band maxima with relative intensity :

667	87	705	45	761	67	803	84	868	77
887	43	924	32	971	20	1099	54	1114	32
1156	65	1169	57	1197	61	1222	70	1242	53
1260	54	1293	78	1309	93	1363	54	1383	69
1417	61	1431	60	1466	59	1532	100	1565	93
1650	99	1754	19	1772	18	2939	34	2987	47
3013	31	3044	24	3074	38	3300	89	3338	93

COMPOUND : **Prosulfocarb**

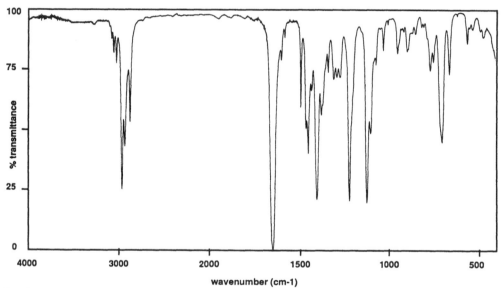

Chemical name : S-benzyl-dipropyl thiocarbamate

Other names : R 15574, SC 0574
Type : herbicide
Brutoformula : C14H21SNO   CAS nr   : -
Molecular mass : 251.38957   Exact mass   : 251.1343766
Instrument : Bruker IFS-85   Optical resolution : 2 cm-1
Scans : 32   Sampling technique : neat film

Band maxima with relative intensity :

565	13	663	26	702	55	769	24	895	16
949	17	1029	15	1123	80	1221	79	1310	27
1342	24	1380	42	1406	78	1454	59	1495	39
1650	100	2874	45	2933	55	2964	74	3029	20
3062	16								

COMPOUND : **Prothoate**

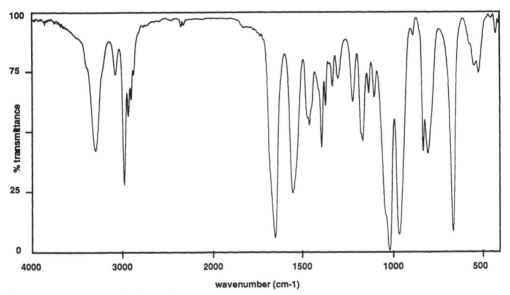

Chemical name : O,O-diethyl S-isopropylcarbamoylmethyl phosphorodithioate

Other names : Fac, Fostion
Type : acaricide, insecticide
Brutoformula : C9H20NO3PS2   CAS nr : 2275-18-5
Molecular mass : 285.35841   Exact mass : 285.0622169
Instrument : Bruker IFS-85   Optical resolution : 2 cm-1
Scans : 32   Sampling technique : neat film

Band maxima with relative intensity :

423	8	517	24	659	92	799	58	827	58
962	93	1013	100	1099	35	1130	33	1163	53
1217	36	1298	27	1329	30	1367	38	1388	56
1456	46	1549	75	1649	94	2934	42	2976	72
3075	25	3292	57						

COMPOUND : **Pyracarbolide**

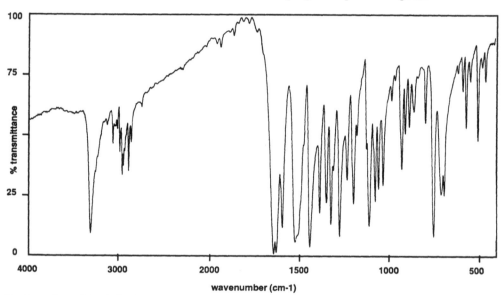

Chemical name : 3,4-dihydro-6-methyl-2H-pyran-5-carboxanilide

Other names   : HOE 2987, Sicarol, W 13764
Type   : fungicide
Brutoformula   : C13H15NO2   CAS nr   : 24691-76-7
Molecular mass : 217.27   Exact mass   : 217.1102704
Instrument   : Bruker IFS-85   Optical resolution : 2 cm-1
Scans   : 32   Sampling technique : KBr pellet

Band maxima with relative intensity :

463	27	507	52	547	27	571	46	590	31
692	75	709	74	752	92	798	44	860	39
884	46	906	49	924	64	980	35	1027	71
1054	72	1074	77	1108	88	1193	78	1229	68
1271	92	1320	87	1346	78	1382	82	1437	97
1521	95	1593	88	1642	100	2853	52	2882	64
2952	66	2979	56	3056	53	3308	91		

344

COMPOUND : **Pyrazophos**

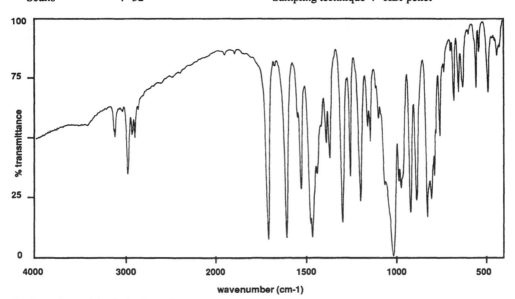

Chemical name	:	O-6-ethoxycarbonyl-5-methylpyrazolo-[1,5-a] pyrimidin-2-yl O,O-diethylphosphorothioate		
Other names	:	W 11099, Afugan, Curamil		
Type	:	fungicide		
Brutoformula	:	C14H20N3O5PS	CAS nr	: 13457-18-6
Molecular mass	:	373.36636	Exact mass	: 373.36636
Instrument	:	Bruker IFS-85	Optical resolution	: 2 cm-1
Scans	:	32	Sampling technique	: KBr pellet

Band maxima with relative intensity :

438	15	487	30	536	13	551	28	624	28
649	30	675	34	753	49	784	66	803	75
827	83	888	76	922	81	973	71	986	68
1020	100	1145	51	1161	45	1202	76	1257	65
1301	85	1370	58	1388	51	1469	91	1530	71
1614	91	1715	92	2907	48	2939	47	2989	64
3126	48								

COMPOUND : **Pyrethrin-I**

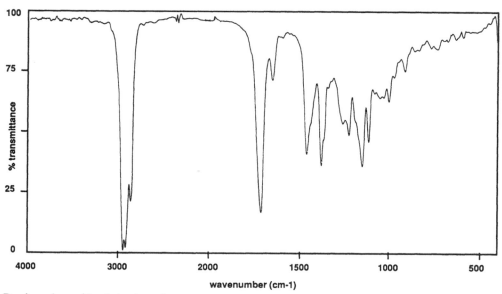

Chemical name   : pyrethrin-I

Other names   : Pyrethrum
Type   : insecticide
Brutoformula   : C21H28O3   CAS nr   : 121-21-1
Molecular mass : 328.45551   Exact mass   : 328.2038311
Instrument   : Bruker IFS-85   Optical resolution : 2 cm-1
Scans   : 32   Sampling technique : neat film

Band maxima with relative intensity :

908	24	995	37	1114	54	1151	65	1224	51
1382	64	1463	59	1652	28	1720	84	1957	15
2869	79	2956	100						

COMPOUND : **Pyridate**

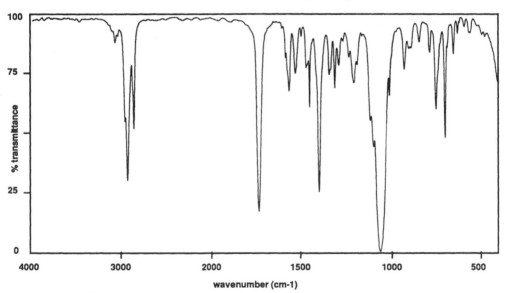

Chemical name	: 6-chloro-3-phenylpyridazin-4-yl S-octylthiocarbonate

Other names   : Fenpyrate, Lentagran
Type   : herbicide
Brutoformula   : C19H23ClN2O2S   CAS nr   : 55512-33-9
Molecular mass : 378.92036   Exact mass   : 378.1168662
Instrument   : Bruker IFS-85   Optical resolution : 2 cm-1
Scans   : 32   Sampling technique : melt

Band maxima with relative intensity :

552	7	626	7	649	16	696	51	745	39
781	15	842	11	925	22	1006	33	1057	100
1202	28	1284	20	1307	30	1338	24	1394.	74
1446	38	1524	24	1560	31	1731	82	2855	47
2926	69	3059	10						

COMPOUND : **Pyridathioben**

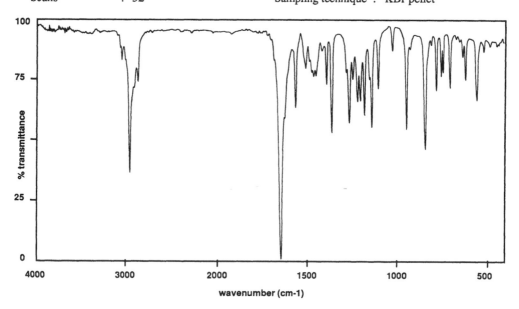

Chemical name   : 4-chloro-2-(1,1-dimethylethyl)-5-[[[4-(1,1dimethylethyl) phenyl] methyl] thio]-3(2H)-pyridazon

Other names   : NC-129
Type   : herbicide

Brutoformula	: C19H25ClN2OS	CAS nr	: 96489-71-3
Molecular mass	: 364.9369	Exact mass	: 364.1376013
Instrument	: Bruker IFS-85	Optical resolution	: 4 cm-1
Scans	: 32	Sampling technique	: KBr pellet

Band maxima with relative intensity :

562	31	622	23	709	27	747	20	758	23
786	29	847	54	949	45	1024	12	1107	28
1144	44	1184	39	1206	33	1222	33	1247	24
1267	42	1366	46	1393	26	1465	23	1511	19
1568	36	1656	100	2866	24	2961	62	3045	16

COMPOUND : **Quaternary Ammonium Comp.**

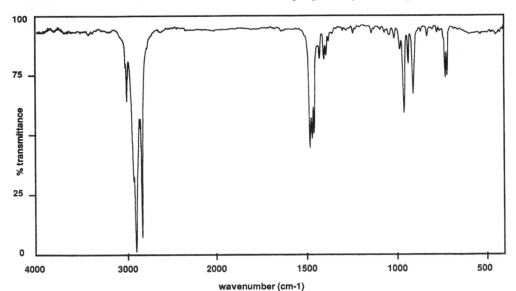

Chemical name   : N-hexyl-N-trimethyl ammonium bromide

Other names   : -
Type   : disinfectant
Brutoformula   : C19H42BrN   CAS nr   : -
Molecular mass   : 364.45729   Exact mass   : 363.2500964
Instrument   : Bruker IFS-85   Optical resolution : 2 cm-1
Scans   : 32   Sampling technique : KBr pellet

Band maxima with relative intensity :

718	37	729	37	910	43	936	32	961	50
1407	30	1461	57	1471	59	1485	62.	2847	94
2916	100	3014	45						

COMPOUND : **Quinalphos**

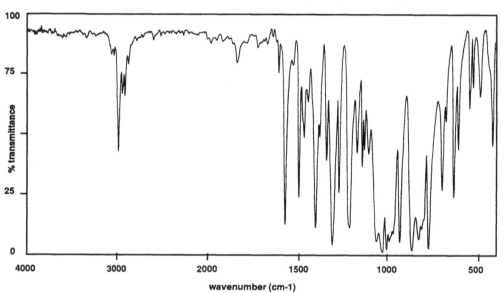

Chemical name : O,O-diethyl O-quinoxalin-2-yl phosphorothioate

Other names : Chinalphos, Bayrusil, Ekalux, Savall
Type : acaricide, insecticide
Brutoformula : C12H15N2O3PS
Molecular mass : 298.29871
Instrument : Bruker IFS-85
Scans : 32

CAS nr : 13593-03-8
Exact mass : 298.0540944
Optical resolution : 2 cm-1
Sampling technique : KBr pellet

Band maxima with relative intensity :

402	18	420	54	487	34	527	30	546	39
606	56	629	76	673	44	694	73	770	98
822	94	860	99	924	95	996	98	1020	100
1100	58	1123	56	1136	63	1165	57	1207	89
1268	74	1306	96	1339	61	1402	89	1445	36
1467	51	1497	76	1574	88	1611	24	1728	13
1839	19	2909	33	2939	32	2982	57		

COMPOUND : **Quinmerac**

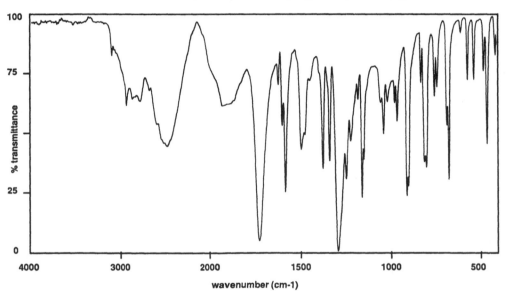

Chemical name	: 3-methyl-7-chloro-8-chinolinic acid		
Other names	: BAS 526		
Type	: insecticide		
Brutoformula	: C11H18NO2Cl	CAS nr	: -
Molecular mass	: 231.102597	Exact mass	: 231.102597
Instrument	: Bruker IFS-85	Optical resolution	: 2 cm-1
Scans	: 32	Sampling technique	: KBr pellet

Band maxima with relative intensity :

407	16	453	54	471	23	525	27	560	27
599	7	663	69	674	46	730	30	743	34
788	64	818	28	899	76	954	44	968	37
1007	36	1028	50	1151	77	1173	35	1214	53
1239	69	1285	100	1330	61	1368	64	1488	56
1576	74	1595	46	1616	29	1723	95	1922	38
2469	55	2766	35	2917	37				

351

COMPOUND : **Quinoclamin**

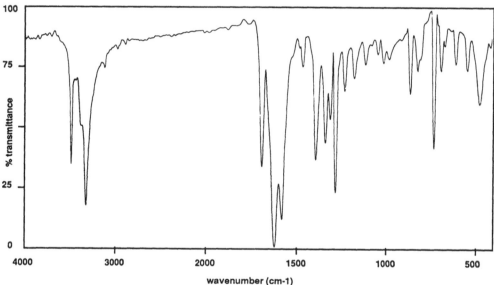

Chemical name   : 2-amino-3-chloro-1,4-naphthachinone

Other names   : Mogeton
Type   : herbicide
Brutoformula   : C10H6ClNO2   CAS nr   : -
Molecular mass : 207.61782   Exact mass   : 207.0087021
Instrument   : Bruker IFS-85   Optical resolution : 4 cm-1
Scans   : 32   Sampling technique : KBr pellet

Band maxima with relative intensity :

466	39	534	25	597	22	680	25	721	58
811	25	852	35	1002	22	1035	18	1107	22
1168	28	1222	33	1274	77	1305	45	1332	55
1384	63	1458	23	1573	88	1614	100	1683	66
3309	82	3462	64						

COMPOUND : **Quintozene**

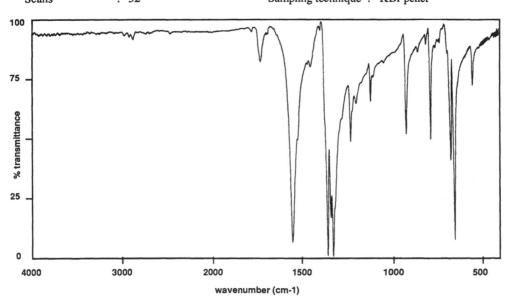

Chemical name	: pentachloronitrobenzene		
Other names	: PCNB, Terrachlor, Tritisan		
Type	: fungicide		
Brutoformula	: C6Cl5NO2	CAS nr	: 82-68-8
Molecular mass	: 295.3374	Exact mass	: 292.8371669
Instrument	: Bruker IFS-85	Optical resolution	: 2 cm-1
Scans	: 32	Sampling technique	: KBr pellet

Band maxima with relative intensity :

552	27	657	93	676	59	787	50	930	48
1121	34	1233	51	1333	100	1345	83	1363	99
1561	93	1731	16						

COMPOUND : **Ofurace**

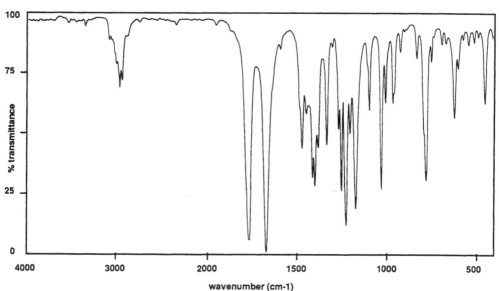

| Chemical name | : | 2-chloro-N-(2,6-dimethylphenyl)-N-(2-oxotetrahydrofuran-3-yl) |
| acetamide | | |

Other names	: RE 20615, Milfuram		
Type	: herbicide		
Brutoformula	: C14H16ClNO3	CAS nr	: 58810-48-3
Molecular mass	: 281.74152	Exact mass	: 281.0818622
Instrument	: Bruker IFS-85	Optical resolution	: 2 cm-1
Scans	: 32	Sampling technique	: KBr pellet

Band maxima with relative intensity :

452	37	542	12	621	43	690	12	749	19
781	70	832	18	922	15	963	37	1002	37
1028	73	1096	40	1173	81	1203	50	1225	88
1251	74	1267	48	1333	54	1381	55	1399	72
1412	68	1471	56	1675	100	1768	94	2953	30

COMPOUND : **Triapenthenol**

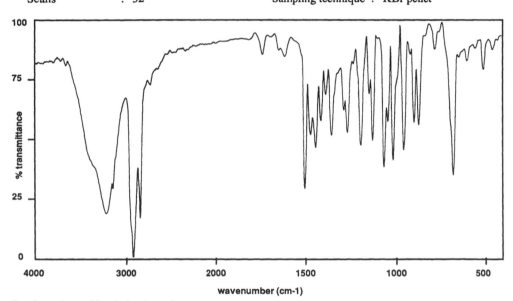

Chemical name   : (E)-1-cyclohexyl-4,4-dimethyl-2-(1,2,4-triazol-1-yl)-1-penten-3-ol

Other names   : Baronet, RSW 0411
Type   : acaricide, fungicide
Brutoformula   : C15H25N3O   CAS nr   : -
Molecular mass : 263.386   Exact mass   : 263.1997487
Instrument   : Bruker IFS-85   Optical resolution : 2 cm-1
Scans   : 32   Sampling technique : KBr pellet

Band maxima with relative intensity :

513	20	678	64	786	11	877	43	904	42
960	54	1018	58	1068	61	1134	50	1155	30
1199	52	1272	47	1361	48	1392	30	1417	41
1446	53	1475	48	1506	70	1623	14	1749	13
2854	83	2927	100	3230	81				

COMPOUND : **Sethoxydim**

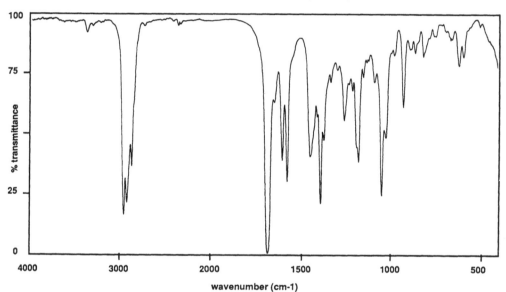

Chemical name	: 2-[1-(ethoxyimino) butyl]-5-[2-(ethylthio) propyl]-3-hydroxy-2-cyclohexen-1-one		
Other names	: NP 55, Poast, Fervinal		
Type	: herbicide		
Brutoformula	: C17H29NO3S	CAS nr	: 74051-80-2
Molecular mass	: 327.48558	Exact mass	: 327.1868016
Instrument	: Bruker IFS-85	Optical resolution	: 2 cm-1
Scans	: 32	Sampling technique	: neat film

Band maxima with relative intensity :

622	20	817	17	862	15	927	38	1053	75
1182	61	1263	43	1398	79	1456	59	1585	69
1612	60	1693	100	2872	62	2928	78	2964	83

COMPOUND : **Simazine**

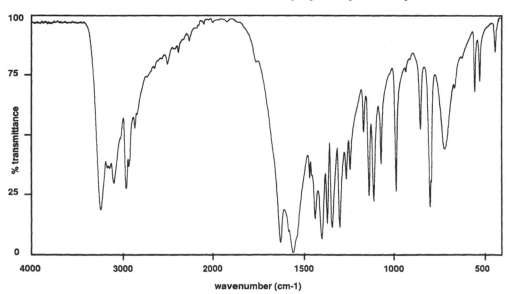

Chemical name	: 2-chloro-4,6-bis(ethylamino)-1,3,5-triazine		
Other names	: Gesatrop, Primatol S, CAT		
Type	: herbicide		
Brutoformula	: C7H12ClN5	CAS nr	: 122-34-9
Molecular mass	: 201.66019	Exact mass	: 210.0781143
Instrument	: Bruker IFS-85	Optical resolution	: 2 cm-1
Scans	: 32	Sampling technique	: KBr pellet

Band maxima with relative intensity :

436	15	522	27	551	31	717	56	799	81
850	47	985	74	1069	62	1110	78	1136	75
1164	48	1243	64	1264	68	1300	89	1345	89
1371	87	1404	94	1440	85	1469	68	1566	100
1635	95	2977	71	3114	69	3261	80		

COMPOUND : **Sodium-dimethyl dithiocarbamate**

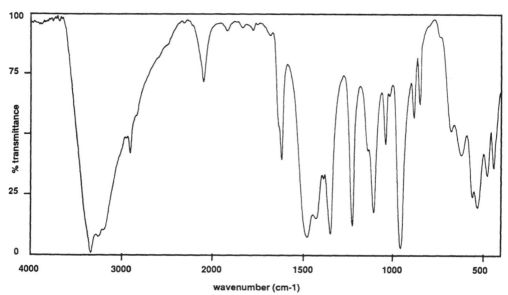

Chemical name	: sodium dimethyl dithiocarbamate (dihydrate)		
Other names	: NaDDC, Neo-Voronit, PP 666, Bayer 22157		
Type	: fungicide, nematicide		
Brutoformula	: C3H6S2NNa.2H2O	CAS nr	: 128-04-1
Molecular mass	: 143.19774	Exact mass	: 142.9839472
Instrument	: Bruker IFS-85	Optical resolution	: 4 cm-1
Scans	: 32	Sampling technique	: KBr pellet

Band maxima with relative intensity :

447	65	484	68	540	82	622	60	848	38
881	44	964	99	1041	55	1116	84	1238	89
1359	92	1488	94	1625	61	2106	28	2923	58
3371	100								

COMPOUND : **Streptomycin (sulphate)**

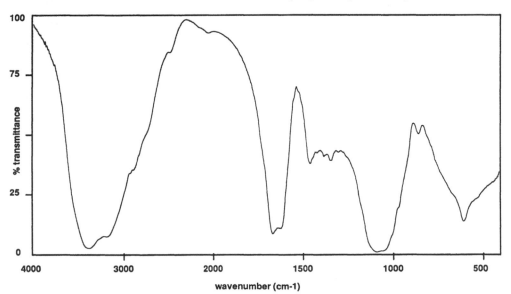

Chemical name	: streptomycin sulphate

Other names	: Fytostrep-60, AAstrepto
Type	: anti bioticum, insecticide
Brutoformula	: C21H39N7O12
Molecular mass	: 581.5846
Instrument	: Bruker IFS-85
Scans	: 32

CAS nr	: 3810-74-0
Exact mass	: 581.26564
Optical resolution	: 2 cm-1
Sampling technique	: KBr pellet

Band maxima with relative intensity :

615	87	1095	100	1463	62	1676	91	3387	96

359

COMPOUND : **Sulfotep**

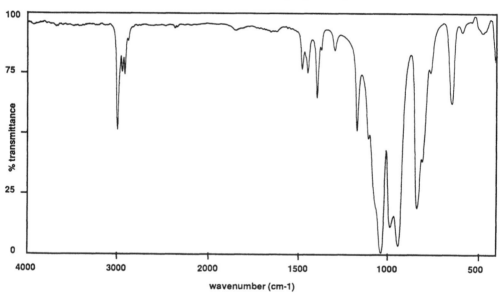

Chemical name	:	O,O,O',O'-tetraethyl dithiopyrophosphate

Other names	:	Bladafum, Thiotepp, Bayer E 393, Dithio			
Type	:	insecticide			
Brutoformula	:	C8H20O5P2S2	CAS nr	:	3689-24-5
Molecular mass	:	322.31312	Exact mass	:	322.0227359
Instrument	:	Bruker IFS-85	Optical resolution	:	2 cm-1
Scans	:	32	Sampling technique	:	neat film

Band maxima with relative intensity :

402	26	472	7	638	36	831	81	932	97
1025	100	1163	48	1292	14	1391	34	1443	23
1475	22	2906	24	2936	23	2984	48		

360

COMPOUND : **2,4,5-T**

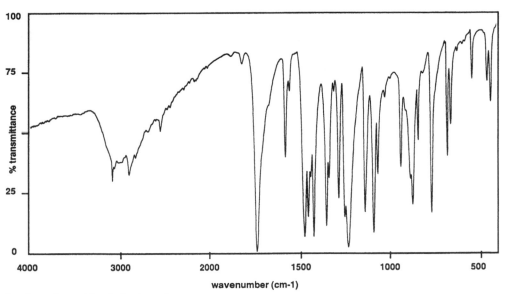

Chemical name   : (2,4,5-trichlorophenoxy) acetic acid

Other names   : Weedone, Trioxone, Transamine
Type   : herbicide
Brutoformula   : C8H5Cl3O3   CAS nr   : 93-76-5
Molecular mass   : 255.48625   Exact mass   : 253.9304246
Instrument   : Bruker IFS-85   Optical resolution   : 2 cm-1
Scans   : 32   Sampling technique : KBr pellet

Band maxima with relative intensity :

442	36	461	27	544	26	660	46	680	59
767	83	840	52	871	80	938	64	1066	67
1089	92	1139	83	1232	98	1287	77	1342	68
1356	89	1429	93	1460	85	1480	93	1587	59
1744	100	2561	48	2915	67	3101	69		

COMPOUND : **2,3,6-TBA**

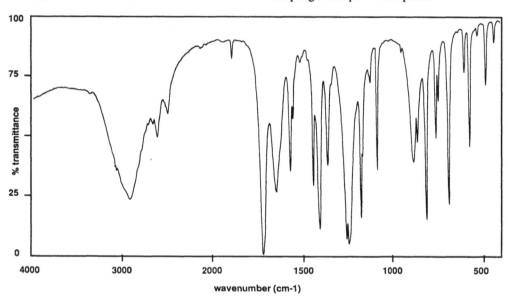

Chemical name   : 2,4,6-trichloro benzoic acid

Other names   : TCB, Trysben, Benzac, HC 1281
Type   : herbicide
Brutoformula   : C7H3Cl3O2   CAS nr   : 50-31-7
Molecular mass   : 225.45976   Exact mass   : 223.9198613
Instrument   : Bruker IFS-85   Optical resolution : 2 cm-1
Scans   : 32   Sampling technique : KBr pellet

Band maxima with relative intensity :

442	11	487	29	574	55	602	24	688	79
745	35	758	51	812	85	860	53	882	61
1087	64	1126	27	1181	84	1252	95	1265	93
1368	62	1413	89	1447	70	1573	64	1651	73
1727	100	1890	16	2493	39	2612	49	2922	75

COMPOUND : **TDE**

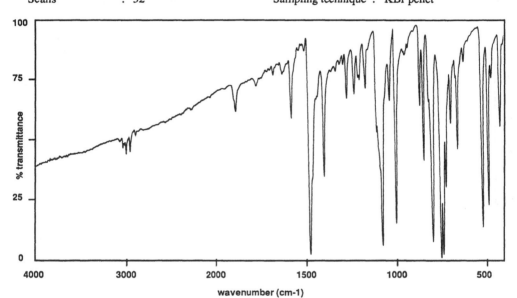

Chemical name   :   1,1-dichloro-2,2-bis (4-chlorophenyl) ethane

Other names   :   4,4'-DDD, Rhothane
Type   :   insecticide
Brutoformula   :   C14H10Cl4   CAS nr   :   72-54-8
Molecular mass   :   320.0478   Exact mass   :   317.953658
Instrument   :   Bruker IFS-85   Optical resolution   :   2 cm-1
Scans   :   32   Sampling technique :   KBr pellet

Band maxima with relative intensity :

432	44	499	77	532	87	632	16	668	53
705	42	734	69	751	98	764	100	810	92
857	58	877	34	1013	84	1044	32	1089	94
1177	27	1211	23	1240	29	1281	31	1411	64
1491	97	1592	39	1900	36	2984	52	3028	53

363

COMPOUND : **Tefluthrin**

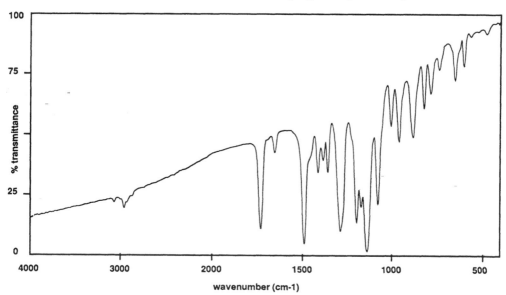

Chemical name	:	2,3,5,6-tetrafluoro-4-methylbenzyl (1RS) cis-3-(Z-2-chloro-3,3,3-trifluoroprop-1-enyl)-2,2-dimethyl cyclopropanecarboxylate		
Other names	:	PP 993		
Type	:	insecticide		
Brutoformula	:	C17H14O2ClF7	CAS nr	: -
Molecular mass	:	418.74173	Exact mass	: 418.0570411
Instrument	:	Bruker IFS-85	Optical resolution	: 2 cm-1
Scans	:	32	Sampling technique	: KBr pellet

Band maxima with relative intensity :

601	21	649	27	738	23	786	33	821	39
881	51	956	53	1001	46	1078	80	1139	100
1197	87	1290	91	1359	66	1386	61	1415	66
1492	96	1656	57	1733	90				

COMPOUND : **Temephos**

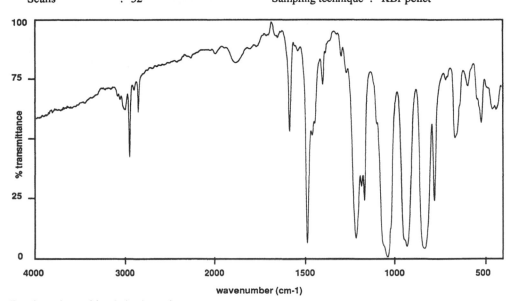

Chemical name : O,O,O',O'-tetramethyl O,O'-thiodi-p-phenylene bis(phosphorothioate)

Other names : Abate, OMS 786, Abathion, Nimitex
Type : insecticide
Brutoformula : C16H20O6P2S3   CAS nr : 3383-96-8
Molecular mass : 466.46172   Exact mass : 465.9897227
Instrument : Bruker IFS-85   Optical resolution : 2 cm-1
Scans : 32   Sampling technique : KBr pellet

Band maxima with relative intensity :

437	37	517	43	594	27	664	49	777	76
834	96	928	95	1033	100	1164	76	1213	92
1399	26	1486	94	1586	46	1884	17	2846	38
2950	57	3002	36						

COMPOUND : **TEPP**

Chemical name	:	O,O,O,O-tetraethyl pyrophosphoric acid

Other names	:	TEPP, Bladan			
Type	:	insecticide			
Brutoformula	:	C8H20P2O7	CAS nr	:	-
Molecular mass	:	290.19192	Exact mass	:	290.0684187
Instrument	:	Bruker IFS-85	Optical resolution	:	2 cm-1
Scans	:	32	Sampling technique	:	neat film

Band maxima with relative intensity :

509	36	756	28	818	41	942	92	980	93
1032	100	1166	52	1296	85	1371	36	1395	36
1445	26	1480	22	1652	10	2914	29	2987	50

COMPOUND : **Terbufos**

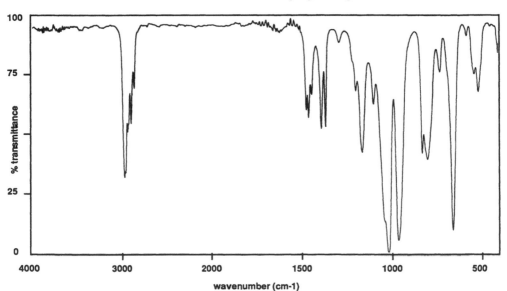

Chemical name : S-tert.butylthiomethyl-O,O-diethyl phosphorodithioate

Other names : Terbuthioate, Counter, AC 92100
Type : insecticide, nematicide
Brutoformula : C9H21O2PS3   CAS nr : 13071-79-9
Molecular mass : 288.42028   Exact mass : 288.0441269
Instrument : Bruker IFS-85   Optical resolution : 2 cm-1
Scans : 32   Sampling technique : neat film

Band maxima with relative intensity :

514	31	657	90	728	23	795	60	827	57
960	94	1014	100	1098	36	1162	57	1366	46
1389	46	1459	42	1471	38	1622	5	2864	28
2899	44	2975	66						

COMPOUND : **Terbutryn**

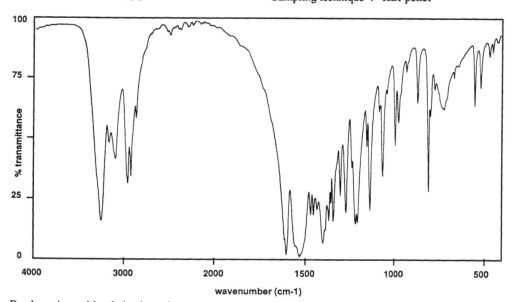

Chemical name	: 2-tert.butylamino-4-ethylamino-6-methylthio-1,3,5-triazine		

Other names	: Igran, GS 14260, Prebane, Clarosan		
Type	: herbicide		
Brutoformula	: C10H19N5S	CAS nr	: 886-5-0
Molecular mass	: 241.35643	Exact mass	: 241.1361061
Instrument	: Bruker IFS-85	Optical resolution	: 2 cm-1
Scans	: 32	Sampling technique	: KBr pellet

Band maxima with relative intensity :

469	15	519	28	553	36	721	37	809	72
865	34	973	43	992	53	1066	65	1137	80
1154	53	1221	86	1274	81	1305	74	1348	84
1372	84	1406	94	1458	82	1474	82	1535	100
1609	99	2930	65	2968	68	3106	57	3266	84

COMPOUND : **Terbutylazin**

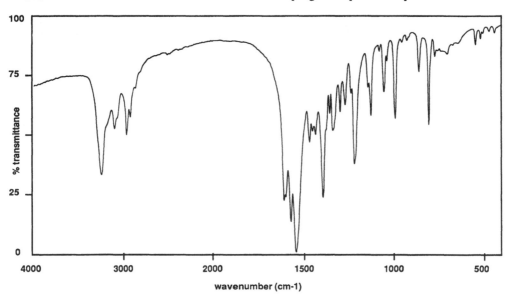

Chemical name   :   2-tert.butylamino-4-ethylamino-6-chloro-1,3,5-triazine

Other names   :   Camparol, GS 13529
Type   :   herbicide
Brutoformula   :   C9H16ClN5   CAS nr   :   5915-41-3
Molecular mass :   229.71437   Exact mass   :   229.1094127
Instrument   :   Bruker IFS-85   Optical resolution : 2 cm-1
Scans   :   32   Sampling technique : KBr pellet

Band maxima with relative intensity :

548	11	772	16	807	45	862	23	993	43
1055	31	1129	41	1223	62	1275	37	1303	40
1345	48	1361	41	1398	76	1474	53	1549	100
1579	87	1618	78	2979	49	3115	47	3260	66

COMPOUND : **Tetrachlorvinphos**

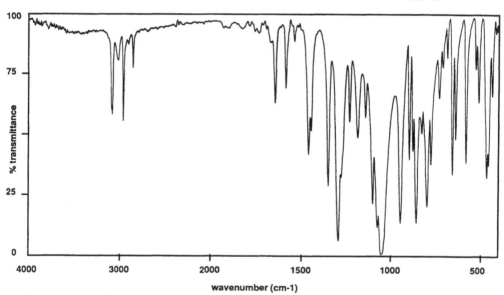

Chemical name : (Z)-2-chloro-1-(2,4,5-trichlorophenyl)vinyl dimethylphosphate

Other names : Rabond, Gardona, Appex, Vinilfos
Type : insecticide, acaricide
Brutoformula : C10H19Cl4O4P   CAS nr   : 22248-79-9
Molecular mass : 365.06659   Exact mass   : 363.8992542
Instrument : Biorad Tracer   Optical resolution : 2 cm-1
Scans : 256   Sampling technique : cryotrapping

Band maxima with relative intensity :

432	33	462	67	507	35	521	21	576	61
633	51	651	66	679	17	704	21	725	34
772	62	795	79	823	48	855	86	871	56
892	59	943	86	1051	100	1099	78	1137	42
1183	50	1228	43	1294	94	1350	70	1443	47
1458	57	1535	10	1582	29	1641	35	2851	20
2958	43	3014	17	3082	40				

COMPOUND : **Tetradifon**

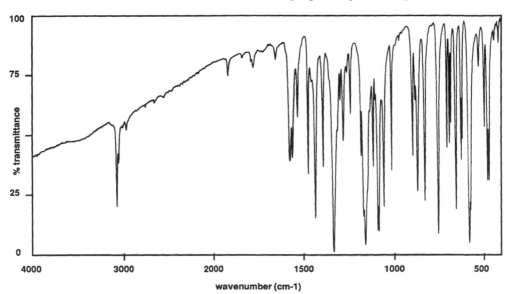

Chemical name   :   4-chlorophenyl-2,4,5-trichlorophenyl sulphone

Other names	: Tedion			
Type	: acaricide			
Brutoformula	: C12H6Cl4O2S	CAS nr	:	116-29-0
Molecular mass	: 356.05242	Exact mass	:	353.8842609
Instrument	: Bruker IFS-85	Optical resolution	:	2 cm-1
Scans	: 32	Sampling technique	:	KBr pellet

Band maxima with relative intensity :

417	11	473	70	481	70	497	47	527	21
583	96	622	52	627	61	656	82	683	45
692	54	706	56	757	92	831	78	872	74
884	35	898	59	1015	65	1059	80	1087	91
1093	90	1117	63	1163	96	1185	59	1243	41
1283	52	1298	35	1338	100	1395	63	1439	85
1478	66	1537	42	1566	59	1580	61	1656	17
1783	21	1920	24	3088	79				

COMPOUND : **Tetraphenyltin**

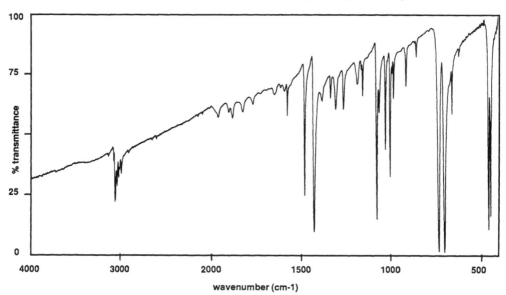

Chemical name   : tetraphenyl tin

Other names   : -
Type   : fungicide
Brutoformula   : C24H20Sn   CAS nr   : 595-90-4
Molecular mass : 427.117   Exact mass   : 428.058692
Instrument   : Bruker IFS-85   Optical resolution : 2 cm-1
Scans   : 32   Sampling technique : KBr pellet

Band maxima with relative intensity :

444	84	456	90	657	41	699	100	730	99
854	17	910	29	976	34	997	67	1024	56
1059	40	1073	85	1150	33	1179	28	1259	39
1302	39	1332	34	1427	90	1479	75	1576	41
1882	42	2985	65	3017	67	3036	70	3060	77

COMPOUND : **Tetramethrin**

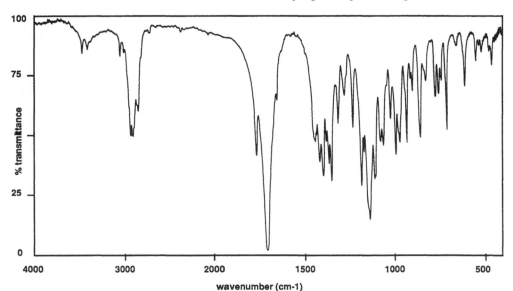

Chemical name : 3,4,5,6-tetrahydrophthalimidomethyl (1RS)-cis/trans-chrysanthemate

Other names : Phthalthrin, Neo-pynamin, SP 1103, OMS 1101
Type : insecticide
Brutoformula : C19H25NO4   CAS nr   : 7696-12-0
Molecular mass : 331.4154   Exact mass   : 331.1783446
Instrument : Bruker IFS-85   Optical resolution : 2 cm-1
Scans : 32   Sampling technique : KBr pellet

Band maxima with relative intensity :

461	22	516	17	548	20	611	31	652	14
711	50	739	29	755	34	775	35	823	29
857	53	899	33	934	55	973	54	996	60
1024	44	1065	56	1080	54	1117	70	1144	87
1176	58	1191	73	1237	48	1283	34	1321	46
1361	70	1375	63	1390	54	1408	68	1427	62
1450	54	1722	100	1778	59	2871	40	2932	50
2958	50	3065	16	3425	13	3483	15		

COMPOUND : **Tetrasul**

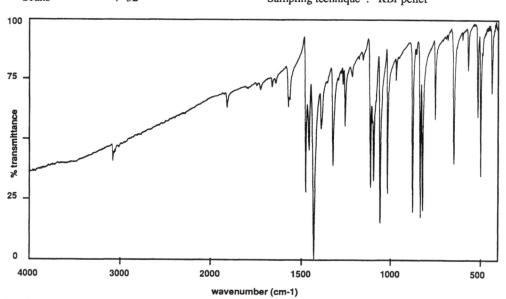

Chemical name   : 2,4,4',5-tetrachlorodiphenyl sulphide

Other names   : V 101, Animert
Type   : acaricide
Brutoformula   : C12H6Cl4S   CAS nr   : 2227-13-6
Molecular mass   : 324.05362   Exact mass   : 321.8944327
Instrument   : Bruker IFS-85   Optical resolution : 2 cm-1
Scans   : 32   Sampling technique : KBr pellet

Band maxima with relative intensity :

434	30	497	65	513	40	567	20	645	59
748	41	819	79	832	82	873	80	963	24
1013	72	1056	84	1092	66	1110	69	1254	43
1266	26	1324	60	1388	44	1431	100	1456	53
1475	71	1570	35	1906	35	3084	58		

374

COMPOUND : **Thiabendazole**

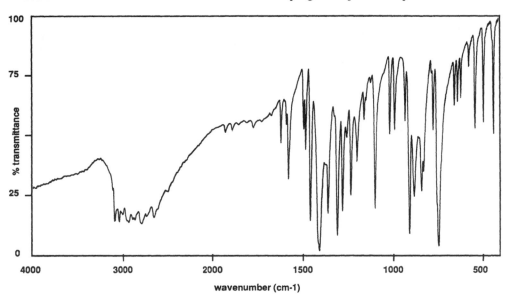

Chemical name   : 2-(thiazol-4-yl) benzimidazole

Other names	: TBZ, Tecto, Comfuval, Mertect		
Type	: fungicide		
Brutoformula	: C10H7N3S	CAS nr	: 148-79-8
Molecular mass	: 201.24739	Exact mass	: 201.0360645
Instrument	: Bruker IFS-85	Optical resolution	: 2 cm-1
Scans	: 32	Sampling technique	: KBr pellet

Band maxima with relative intensity :

432	50	489	45	537	48	570	22	617	35
636	37	652	38	739	98	769	49	821	66
833	75	875	77	902	93	926	45	986	48
1012	50	1095	82	1157	44	1196	62	1231	76
1252	52	1277	83	1305	93	1357	84	1404	100
1455	87	1481	57	1491	48	1579	69	1591	46
1622	54	1776	47	1929	49	2653	85	2793	88
2935	87	3043	87	3096	87				

375

COMPOUND : **Thiameturon**

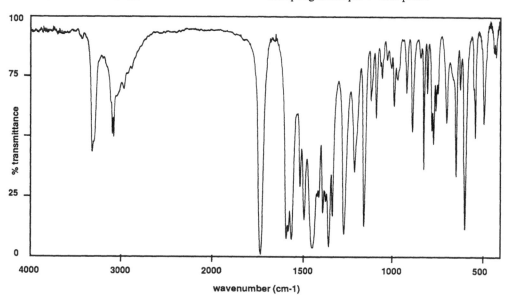

Chemical name : methyl-3-[[[[(4-methoxy-6-methyl-1,3,5-triazin-2-yl) amino] carbonyl]
amino] sulphonyl]-2-thiophenecarboxylate
Other names : DPX-M6316, INM 6316
Type : herbicide
Brutoformula : C12H13S2N5O6   CAS nr   : -
Molecular mass : 387.38731   Exact mass   : 387.0307158
Instrument : Bruker IFS-85   Optical resolution : 2 cm-1
Scans : 32   Sampling technique : KBr pellet

Band maxima with relative intensity :

422	16	487	44	535	50	590	89	617	30
637	66	690	44	751	39	763	53	773	47
799	31	817	63	878	48	911	31	980	37
1048	25	1082	42	1112	34	1150	88	1204	64
1265	91	1330	83	1352	96	1384	82	1441	97
1488	85	1512	71	1559	93	1588	93	1731	100
3076	50	3092	49	3317	56				

COMPOUND : **Thiofanox**

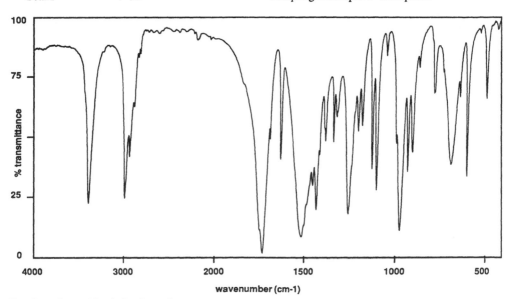

Chemical name : 3,3-dimethyl-1-(methylthio) butanone-O-methylcarbamoyloxime

Other names : Dacamox, DS 15647, ENT 27851
Type : acaricide, insecticide
Brutoformula : C9H18N2O2S   CAS nr   : 39196-18-4
Molecular mass : 218.31601   Exact mass   : 218.1088901
Instrument : Bruker IFS-85   Optical resolution   : 2 cm-1
Scans : 32   Sampling technique : KBr pellet

Band maxima with relative intensity :

474	32	582	66	622	32	670	61	767	31
848	20	885	56	911	64	951	89	973	52
1028	15	1080	72	1106	63	1162	45	1184	47
1235	82	1301	41	1319	52	1363	52	1412	81
1433	70	1494	92	1618	59	1681	51	1716	100
2918	59	2967	70	3377	79				

COMPOUND : **Thiometon**

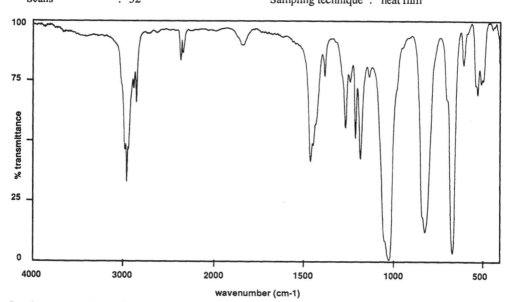

Chemical name   : S-2-ethylthioethyl-O,O-dimethylphosphorodithioate

Other names	: Ekatin, S2030, Ebicid			
Type	: acaricide, insecticide			
Brutoformula	: C6H15O2PS3	CAS nr	: 640-15-3	
Molecular mass	: 246.33901	Exact mass	: 245.9971793	
Instrument	: Bruker IFS-85	Optical resolution	: 2 cm-1	
Scans	: 32	Sampling technique	: neat film	

Band maxima with relative intensity :

499	25	520	30	600	17	658	97	814	88
1015	100	1176	57	1205	48	1260	44	1376	22
1454	58	1832	9	2840	33	2945	66		

COMPOUND : **Thiophanate-methyl**

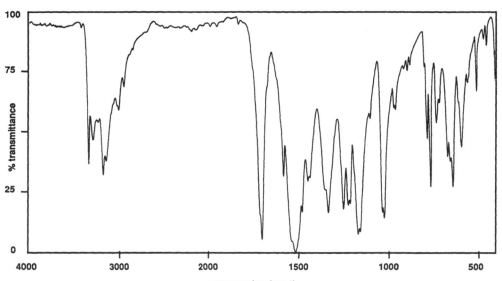

Chemical name   : dimethyl-4,4'-(o-phenylene) bis-(3-thioallophanate)

Other names   : Topsin-M, BET-methyl, Cercobin-M, NF 44
Type   : acaricide, fungicide
Brutoformula   : C12H14N4O4S2   CAS nr   : 23564-05-8
Molecular mass   : 342.38978   Exact mass   : 342.045639
Instrument   : Bruker IFS-85   Optical resolution : 2 cm-1
Scans   : 32   Sampling technique : KBr pellet

Band maxima with relative intensity :

450	13	508	32	594	56	642	73	670	60
730	45	764	73	782	52	960	40	1027	86
1173	92	1227	80	1254	82	1339	83	1452	70
1522	100	1589	68	1710	94	3015	39	3194	66
3301	51	3350	61						

COMPOUND : **Thiram**

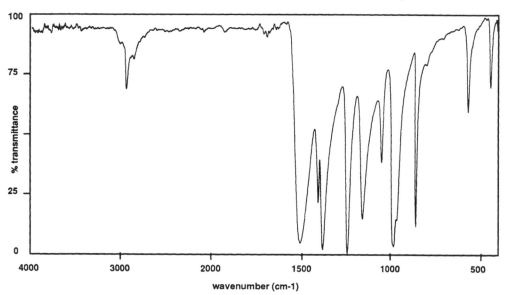

Chemical name : tetramethylthiouram disulphide

Other names   : TMTD, Thiuram, Arasan, Tersan
Type   : fungicide
Brutoformula   : C6H12N2S4   CAS nr   : 137-26-8
Molecular mass : 240.41594   Exact mass   : 239.9883324
Instrument   : Bruker IFS-85   Optical resolution : 2 cm-1
Scans   : 32   Sampling technique : KBr pellet

Band maxima with relative intensity :

440	29	562	39	847	88	969	97	1038	61
1148	85	1235	100	1374	98	1400	78	1501	95
1690	8	2927	30						

COMPOUND : **Tolclophos-methyl**

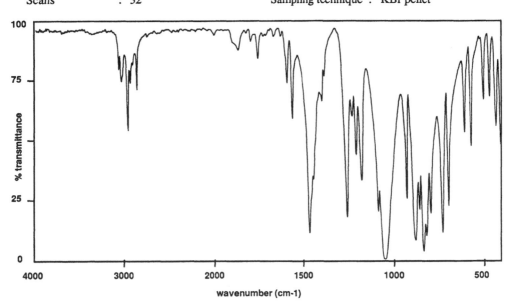

Chemical name   : O-2,6-dichloro-p-tolyl O,O-dimethyl phosphorothioate

Other names   : Salithion, S 3349, Rizolex, AA 3685
Type   : fungicide
Brutoformula   : C9H11Cl2O3PS   CAS nr   : 57018-04-9
Molecular mass : 301.12598   Exact mass   : 299.9543558
Instrument   : Bruker IFS-85   Optical resolution   : 2 cm-1
Scans   : 32   Sampling technique : KBr pellet

Band maxima with relative intensity :

430	43	463	31	499	32	571	52	609	46
701	77	735	88	798	80	840	96	862	79
885	92	930	74	1050	100	1087	79	1178	66
1208	55	1228	39	1259	81	1469	88	1564	39
1592	24	1756	14	1866	10	2851	26	2928	23
2957	43	3024	22	3052	17				

COMPOUND : **Toluenesulphonchloramide (sodium salt)**

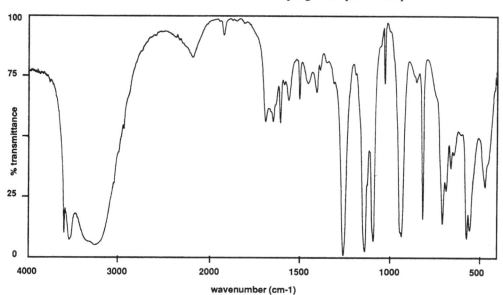

Chemical name   : sodium p-toluenesulphonchloramide

Other names   : Chloramine-T, tosylchloroamide-Na

Type	:			
Brutoformula	: C7H7ClNO2SNa	CAS nr	:	-
Molecular mass	: 227.64211	Exact mass	:	226.9783804
Instrument	: Bruker IFS-85	Optical resolution	:	2 cm-1
Scans	: 128	Sampling technique	:	KBr pellet

Band maxima with relative intensity :

462	71	549	89	566	92	650	63	698	86
809	84	927	91	1018	27	1085	93	1133	98
1252	100	1398	31	1493	33	1553	34	1599	43
1640	43	1683	43	1917	6	2179	16	3254	95
3530	93	3588	90						

COMPOUND : **Tolylfluanid**

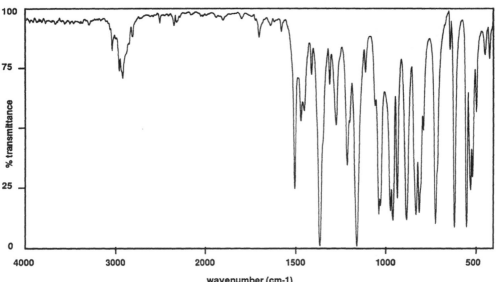

Chemical name   : N-dichlorofluoromethylthio-N,N'-dimethyl-N-p-tolylsulphamide

Other names   : Euparen-M
Type   : acaricide, fungicide
Brutoformula   : C10H13Cl2FN2O2S2   CAS nr   : 731-27-1
Molecular mass   : 347.25171   Exact mass   : 345.9779482
Instrument   : Bruker IFS-85   Optical resolution : 2 cm-1
Scans   : 32   Sampling technique : KBr pellet

Band maxima with relative intensity :

422	20	447	18	495	43	516	70	528	76
550	91	618	92	640	16	722	90	789	50
814	85	832	86	885	88	935	79	958	89
972	84	1037	86	1112	25	1160	100	1214	65
1275	48	1311	31	1366	99	1412	27	1472	46
1507	75	1583	8	1709	11	2928	28	2968	25
3046	16								

383

COMPOUND : **Triadimefon**

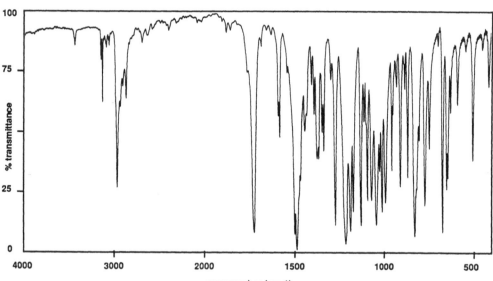

Chemical name   : 1-(4-chlorophenoxy)-3,3-dimethyl-1-(1,2,4-triazol-1-yl) butanone

Other names	: Bayleton, MEB 6447, Amiral		
Type	: fungicide		
Brutoformula	: C14H16ClN3O2	CAS nr	: 43121-43-3
Molecular mass	: 293.75552	Exact mass	: 293.0930945
Instrument	: Bruker IFS-85	Optical resolution	: 2 cm-1
Scans	: 32	Sampling technique	: KBr pellet

Band maxima with relative intensity :

420	31	507	62	590	39	628	42	643	69
649	74	674	92	746	57	773	81	804	54
830	94	867	69	883	32	909	73	956	66
992	80	1011	84	1026	66	1046	89	1074	79
1096	79	1115	45	1133	89	1177	83	1192	92
1220	97	1277	89	1342	58	1351	50	1379	61
1396	42	1448	49	1490	100	1502	93	1585	52
1594	43	1726	92	2872	35	2974	72	3139	36

384

COMPOUND : **Triadimenol**

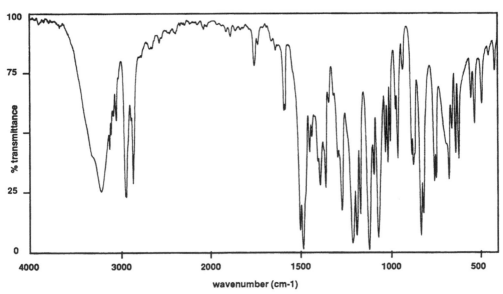

Chemical name : 1-(4-chlorophenoxy)-3,3-dimethyl-1-(1H-1,2,4-triazol-1-yl)-butan-2-ol

Other names : Baytan, Bayfidan, KWG 0519
Type : fungicide
Brutoformula : C14H18ClN3O2   CAS nr   : 55219-65-3
Molecular mass : 295.77146   Exact mass   : 295.1087437
Instrument : Bruker IFS-85   Optical resolution : 2 cm-1
Scans : 32   Sampling technique : KBr pellet

Band maxima with relative intensity :

420	24	493	38	534	46	553	35	621	61
637	59	657	48	676	70	748	69	757	71
818	84	832	94	872	64	883	59	932	23
964	61	975	40	1005	54	1020	62	1034	58
1074	94	1097	68	1126	100	1175	84	1194	93
1218	97	1275	83	1365	73	1395	70	1453	61
1490	98	1507	92	1595	40	2882	60	2959	64
3237	62								

COMPOUND : **Triallate**

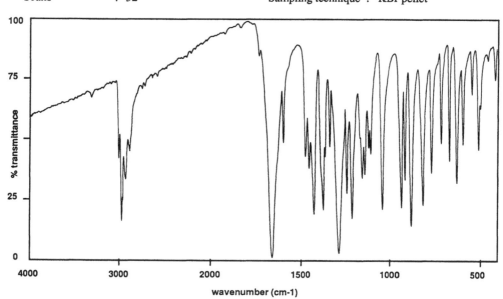

Chemical name   : S-2,3,3-trichloroallyl diisopropylthiocarbamate

Other names   : Avadex, CP 23426, Far-Go
Type   : herbicide
Brutoformula   : C10H16Cl3NOS   CAS nr   : 2303-17-5
Molecular mass : 304.66412   Exact mass   : 303.0018129
Instrument   : Bruker IFS-85   Optical resolution : 2 cm-1
Scans   : 32   Sampling technique : KBr pellet

Band maxima with relative intensity :

505	55	541	31	591	52	624	69	665	59
711	52	766	64	814	78	880	87	911	67
932	79	1039	79	1104	56	1114	53	1136	65
1152	66	1211	83	1239	72	1287	98	1333	53
1373	79	1425	81	1450	62	1471	57	1595	51
1664	100	2926	66	2972	83	3003	57		

COMPOUND : **Triamiphos**

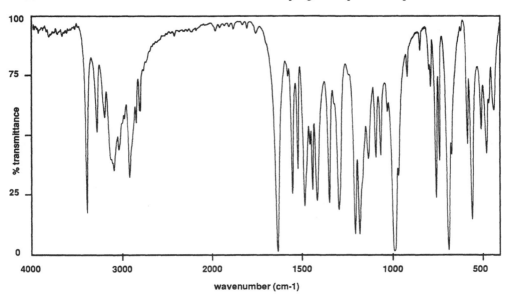

Chemical name : P-5-amino-3-phenyl-1,2,4-triazol-1-yl-N,N,N',N'-tetramethyl-
phosphonic diamide
Other names : Wepesyn, WP 155
Type : acaricide, insecticide, fungicide
Brutoformula : C12H19N6OP   CAS nr   : 1031-47-6
Molecular mass : 294.29859   Exact mass   : 294.1357847
Instrument : Bruker IFS-85   Optical resolution : 2 cm-1
Scans : 32   Sampling technique : KBr pellet

Band maxima with relative intensity :

438	40	480	58	508	48	561	86	584	54
692	99	740	61	760	77	786	30	844	14
918	25	995	100	1067	59	1095	60	1136	60
1188	92	1212	92	1301	82	1353	79	1421	78
1446	73	1459	54	1489	80	1527	64	1559	75
1646	99	2810	39	2856	44	2934	67	3052	55
3107	64	3208	41	3292	47	3405	82		

COMPOUND : **Triazophos**

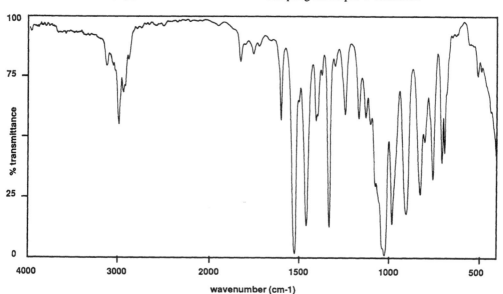

Chemical name : O,O-diethyl-O-1-phenyl-1,2,4-triazol-3-yl-phosphorothioate

Other names : Hostathion, HOE 2960, WL 12712
Type : acaricide, fungicide, insecticide
Brutoformula : C12H16N3O3PS   CAS nr   : 24017-47-8
Molecular mass : 313.31338   Exact mass   : 313.0649922
Instrument : Bruker IFS-85   Optical resolution : 2 cm-1
Scans : 32   Sampling technique : neat film

Band maxima with relative intensity :

504	24	690	56	705	60	756	67	828	74
904	82	979	86	1024	100	1124	41	1163	42
1238	40	1330	87	1402	43	1459	87	1526	99
1599	42	1754	14	1825	17	2931	30	2982	44
3119									

COMPOUND   : **Trichlofenidin**

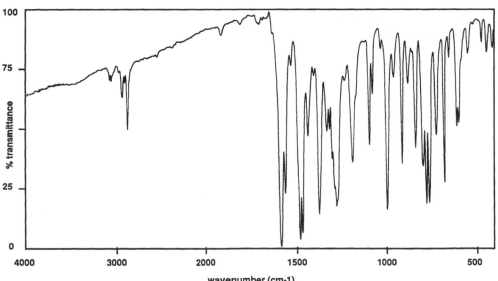

Chemical name   : N,N'-di (3-chloro)phenyl-2-trichloromethyl-perhydro-1,3-pyrazoline

Other names   : UTH 1412, OMS 1824
Type   : insecticide
Brutoformula   : C16H13Cl5N2   CAS nr   :
Molecular mass   : 410.56041   Exact mass   : 407.9521317
Instrument   : Bruker IFS-85   Optical resolution : 2 cm-1
Scans   : 32   Sampling technique : KBr pellet

Band maxima with relative intensity :

410	15	441	17	470	13	546	18	598	47
609	49	649	19	679	73	725	52	763	81
779	81	797	66	838	58	878	30	913	65
958	28	998	84	1080	35	1098	56	1193	64
1283	82	1322	50	1336	51	1379	86	1440	53
1471	94	1485	97	1568	77	1591	100	2888	49
2943	35	3065	28						

COMPOUND : **Trichloro-phenyltin**

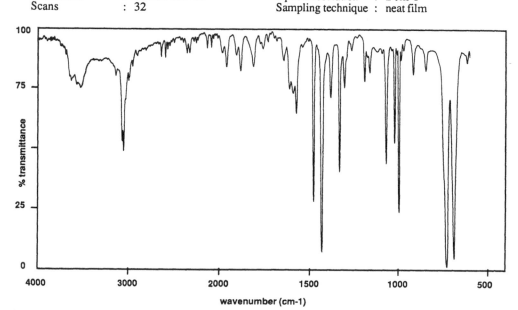

Chemical name   : trichloro-phenyltin

Other names   :
Type   : fungicide
Brutoformula   : C6H5Cl3Sn   CAS nr   : 1124-19-2
Molecular mass   : 248.8457   Exact mass   : 245.87481
Instrument   : Bruker IFS-85   Optical resolution : 2 cm-1
Scans   : 32   Sampling technique : neat film

Band maxima with relative intensity :

688	96	730	100	846	17	916	18	996	76
1020	47	1067	55	1163	17	1190	21	1305	23
1333	59	1382	27	1433	93	1479	71	1574	34
1645	12	1759	7	1812	14	1882	16	1959	14
3057	49	3073	47						

COMPOUND : **Trichlorfon**

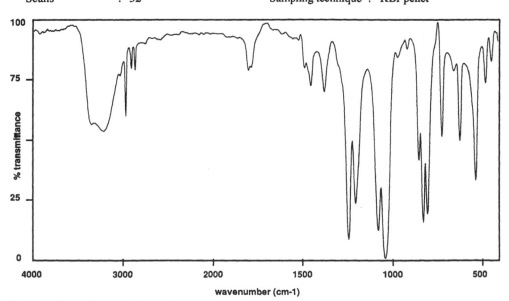

Chemical name   : dimethyl-(2,2,2-trichloro-1-hydroxyethyl) phosphonate

Other names   : Dipterex, Neguvan, Chlorofos, Metriphonate
Type   : insecticide
Brutoformula   : C4H8Cl3O4P   CAS nr   : 52-68-6
Molecular mass : 257.43872   Exact mass   : 255.9225765
Instrument   : Bruker IFS-85   Optical resolution : 2 cm-1
Scans   : 32   Sampling technique : KBr pellet

Band maxima with relative intensity :

443	16	476	26	532	67	621	50	719	48
800	81	825	84	848	58	1039	100	1078	88
1201	76	1242	91	1373	29	1450	26	1799	19
2858	19	2898	18	2962	38	3209	45		

391

COMPOUND : **Trichloronate**

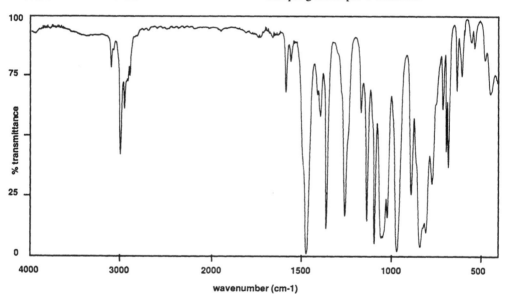

Chemical name   : O-ethyl-O-(2,4,5-trichlorophenyl) ethylphosphonothioate

Other names   : Phytosol, Agritox, S 4400, Agrisil
Type   : insecticide
Brutoformula   : C10H12Cl3O2PS   CAS nr   : 327-98-0
Molecular mass   : 333.5987   Exact mass   : 331.9361194
Instrument   : Bruker IFS-85   Optical resolution   : 2 cm-1
Scans   : 32   Sampling technique : neat film

Band maxima with relative intensity :

441	31	531	12	601	24	628	30	674	62
685	56	705	38	765	69	829	96	878	74
954	98	1009	84	1044	93	1082	95	1125	85
1160	40	1245	83	1349	89	1384	41	1459	100
1555	18	1580	31	1958	20	2939	38	2980	58
3092	21								

COMPOUND : **2,4,5-Trichlorophenol**

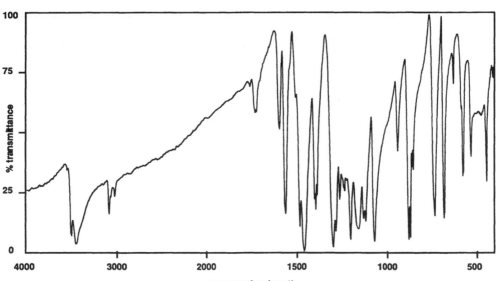

Chemical name	: 2,4,5-trichlorophenol		
Other names	: TCP		
Type	: fungicide		
Brutoformula	: C6H3Cl3O	CAS nr	: 95-95-4
Molecular mass	: 197.4492	Exact mass	: 195.924947
Instrument	: Bruker IFS-85	Optical resolution	: 2 cm-1
Scans	: 32	Sampling technique	: KBr pellet

Band maxima with relative intensity :

441	71	527	60	573	68	622	29	680	86
730	85	849	65	869	94	879	95	935	57
1071	96	1120	87	1161	90	1205	95	1235	74
1263	78	1300	98	1393	82	1459	100	1483	89
1564	84	1595	48	1729	41	3090	83	3454	95
3499	92								

COMPOUND : **Triclopyr**

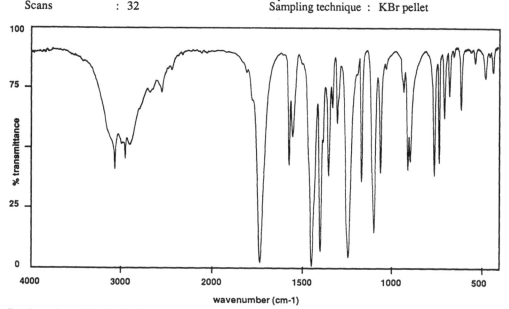

Chemical name	: 3,5,6-trichloro-2-pyridyloxyacetic acid		
Other names	: Dowco 233, Garlon		
Type	: herbicide		
Brutoformula	: C7H4Cl3NO3	CAS nr	: 55335-06-3
Molecular mass	: 256.47383	Exact mass	: 254.9256732
Instrument	: Bruker IFS-85	Optical resolution	: 2 cm-1
Scans	: 32	Sampling technique	: KBr pellet

Band maxima with relative intensity :

433	18	476	20	533	14	608	33	673	28
702	37	732	56	759	61	892	55	905	59
927	26	1055	60	1094	85	1164	64	1240	96
1302	39	1328	32	1351	61	1398	93	1445	100
1553	45	1572	57	1736	98	2950	54	3068	58

COMPOUND : **Tridemorph**

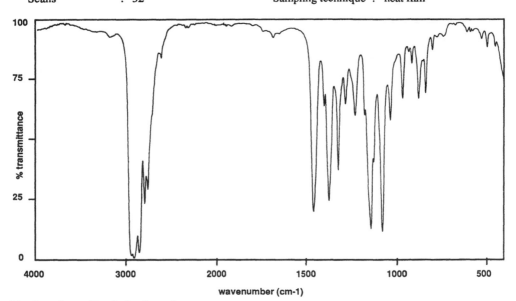

Chemical name : 2,6-dimethyl-4-tridecylmorpholine

Other names	: Calixin, BAS 2201F		
Type	: fungicide		
Brutoformula	: C19H39NO	CAS nr	: 24602-86-6
Molecular mass	: 297.52878	Exact mass	: 297.3031467
Instrument	: Bruker IFS-85	Optical resolution	: 2 cm-1
Scans	: 32	Sampling technique	: neat film

Band maxima with relative intensity :

493	10	798	11	838	30	878	32	916	17
967	32	1035	41	1082	89	1145	88	1229	39
1283	34	1323	63	1376	76	1401	35	1461	80
1958	6	2773	70	2809	76	2869	97	2926	100

COMPOUND : **Trifluralin**

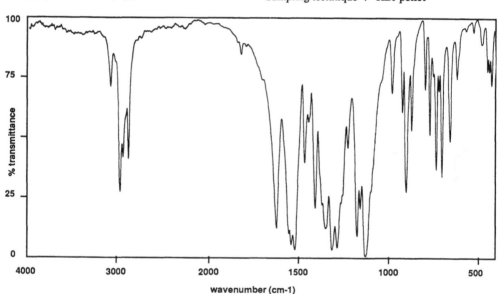

Chemical name   : α,α,α-trifluoro-2,6-dinitro-N,N-dipropyl-p-toluidine

Other names   : Treflan, Digermin, Trefanocide
Type   : herbicide
Brutoformula   : C13H16F3N3O4   CAS nr   : 1582-09-8
Molecular mass : 335.28537   Exact mass   : 335.1092762
Instrument   : Bruker IFS-85   Optical resolution : 2 cm-1
Scans   : 32   Sampling technique : KBr pellet

Band maxima with relative intensity :

428	28	450	22	482	10	621	25	662	51
709	66	724	30	740	63	776	49	800	29
874	47	905	73	925	39	980	31	1133	100
1161	80	1178	91	1228	54	1287	96	1317	97
1352	88	1410	79	1469	60	1526	97	1630	88
2880	58	2941	57	2973	72	3081	27		

COMPOUND : **Triforine**

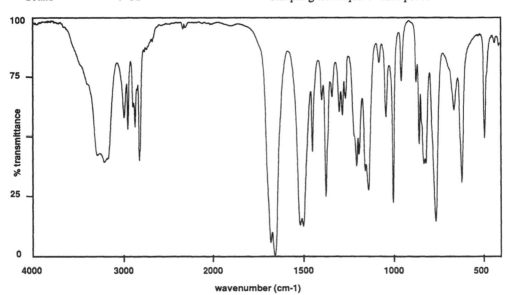

Cl₃C—CH—NH—CHO

Cl₃C—CH—NH—CHO

Chemical name	: 1,4-bis (2,2,2-trichloro-1-formamidoethyl) piperazine		
Other names	: Funginex, Saprol		
Type	: fungicide		
Brutoformula	: C10H14Cl6N4O2	CAS nr	: 26644-46-2
Molecular mass	: 434.96668	Exact mass	: 431.924784
Instrument	: Bruker IFS-85	Optical resolution	: 2 cm-1
Scans	: 32	Sampling technique	: KBr pellet

Band maxima with relative intensity :

495	50	620	69	664	38	765	85	832	61
858	52	876	26	958	25	1004	77	1045	41
1083	18	1142	72	1194	57	1208	62	1270	33
1287	40	1304	39	1345	32	1379	75	1402	34
1454	55	1503	87	1661	100	2836	59	2884	45
2967	46	3008	41	3226	60				

COMPOUND : **Vamidothion**

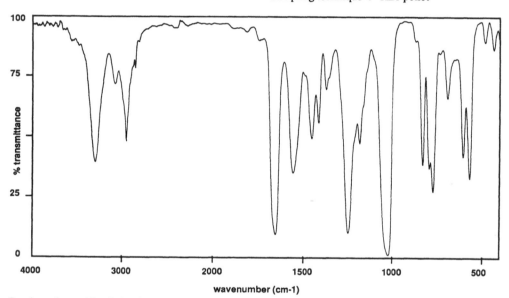

Chemical name   : O,O-dimethyl-S-2-(1-methylcarbamoylethylthio) ethyl phosphorothioate

Other names	: Vation, Vatox, Kilval, Trucidor		
Type	: acaricide, insecticide		
Brutoformula	: C8H18NO4PS2	CAS nr	: 2275-23-2
Molecular mass	: 287.33072	Exact mass	: 287.0414818
Instrument	: Bruker IFS-85	Optical resolution	: 2 cm-1
Scans	: 32	Sampling technique	: KBr pellet

Band maxima with relative intensity :

430	13	478	10	567	68	603	58	688	33
773	73	829	61	1022	100	1184	52	1249	90
1371	29	1411	43	1450	50	1556	65	1656	91
2952	51	3082	26	3305	59				

COMPOUND : **Vernolate**

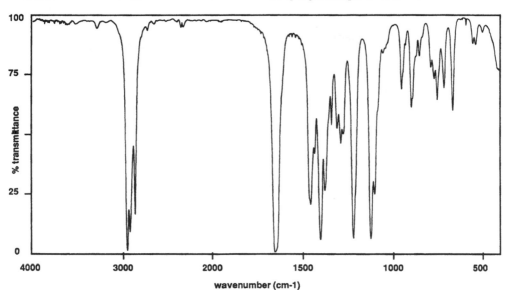

Chemical name   : S-propyl dipropylthiocarbamate

Other names   : Vernam, Surpass
Type   : herbicide
Brutoformula   : C10H21NOS   CAS nr   : 1929-77-7
Molecular mass   : 203.3449   Exact mass   : 203.1343766
Instrument   : Bruker IFS-85   Optical resolution : 2 cm-1
Scans   : 32   Sampling technique : neat film

Band maxima with relative intensity :

536	11	665	39	714	30	751	35	849	16
895	38	950	30	1101	75	1122	94	1220	94
1290	53	1311	47	1342	45	1379	73	1405	94
1459	79	1656	100	2874	83	2933	90	2962	98

COMPOUND : **Vinclozolin**

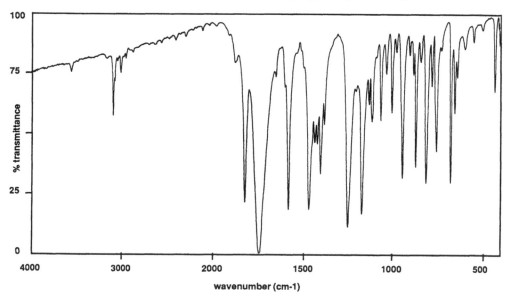

Chemical name : 3-(3,3-dichlorophenyl)-5-methyl-5-vinyloxazolidine-2,4-dione

Other names : Ronilan, Vorlan, Ornalin, BAS 352F
Type : fungicide
Brutoformula : C12H9Cl2O3   CAS nr : 50471-44-8
Molecular mass : 286.11643   Exact mass : 284.9959431
Instrument : Bruker IFS-85   Optical resolution : 2 cm-1
Scans : 32   Sampling technique : KBr pellet

Band maxima with relative intensity :

427	31	544	10	592	13	632	25	646	40
668	70	746	56	772	29	803	70	833	19
859	63	872	24	893	16	930	68	965	15
990	40	1019	24	1052	43	1102	44	1119	37
1158	83	1238	88	1373	45	1394	66	1412	53
1427	53	1459	81	1574	81	1736	100	1815	78
2997	23	3085	41						

COMPOUND : **Zineb**

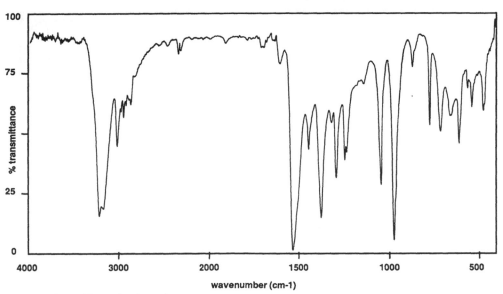

(x > 1)

Chemical name	:	zinc ethylenebis(dithiocarbamate)

Other names	:	Dithane Z-78, Phytox, Parzate, Zinosan			
Type	:	fungicide			
Brutoformula	:	C4H6N2S4Zn	CAS nr	:	12122-67-7
Molecular mass	:	275.72582	Exact mass	:	273.8705258
Instrument	:	Bruker IFS-85	Optical resolution	:	2 cm-1
Scans	:	32	Sampling technique	:	KBr pellet

Band maxima with relative intensity :

477	41	541	39	614	55	660	43	717	50
777	47	874	22	976	95	1048	72	1245	61
1294	69	1379	86	1446	57	1538	100	1608	20
1715	13	2959	43	3030	55	3230	85		

COMPOUND : **Ziram**

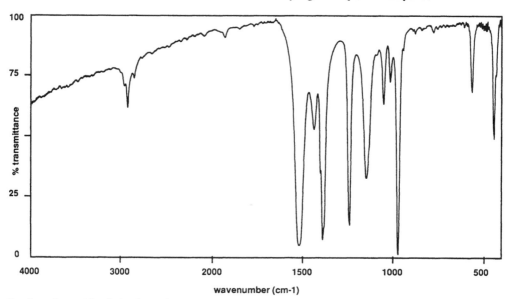

Chemical name   : zinc dimethyldithiocarbamate

Other names   : Milban, AAprotect, Fuklasin
Type   : fungicide, repellent
Brutoformula   : C6H12N2S4Zn   CAS nr   : 2275-23-2
Molecular mass : 305.7959   Exact mass   :303.9174734
Instrument   : Bruker IFS-85   Optical resolution : 2 cm-1
Scans   : 32   Sampling technique : KBr pellet

Band maxima with relative intensity :

444	51	565	31	973	100	1013	26	1050	36
1149	67	1243	87	1391	93	1438	47	1522	96
2927	37								

COMPOUND : **2',4-DDD**

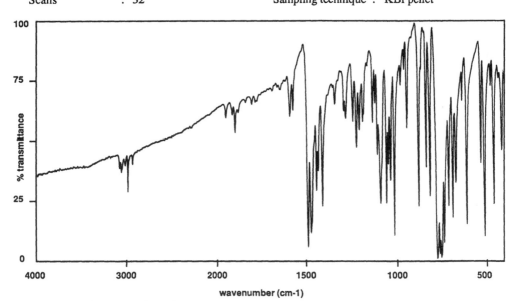

Chemical name : 2',4-dichloro-1,1-diphenyl-2,2-dichloroethane

Other names : DDD
Type : metabolite of DDT, pesticide
Brutoformula : C14H10Cl4   CAS nr : 53-19-0
Molecular mass : 320.0478   Exact mass : 317.953658
Instrument : Bruker IFS-85   Optical resolution : 2 cm-1
Scans : 32   Sampling technique : KBr pellet

Band maxima with relative intensity :

459	77	510	90	529	59	609	85	668	67
686	82	708	77	734	93	752	99	758	97
773	100	814	73	835	61	878	77	944	44
1015	89	1034	66	1046	59	1056	76	1089	76
1107	55	1118	33	1133	42	1187	41	1205	45
1221	52	1240	41	1278	40	1289	36	1339	33
1410	77	1433	62	1443	70	1473	88	1490	94
1571	36	1591	38	1901	45	1916	38	1952	39
3066	78								

403

COMPOUND : **2',4-DDE**

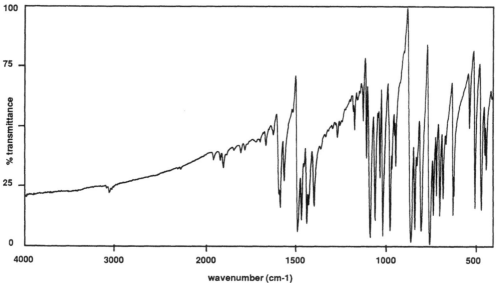

Chemical name   : 2',4-dichloro-1,1-diphenyl-2,2-dichloroethene

Other names   : DDE
Type   : metabolite of DDT, pesticide
Brutoformula   : C14H8Cl4   CAS nr   : 3424-82-6
Molecular mass : 318.03186   Exact mass   : 315.938009
Instrument   : Bruker IFS-85   Optical resolution : 2 cm-1
Scans   : 32   Sampling technique : KBr pellet

Band maxima with relative intensity :

440	68	447	62	468	85	503	84	534	50
626	87	681	80	700	87	718	79	735	87
757	100	805	94	839	93	862	99	944	66
949	53	963	67	974	94	1014	96	1030	71
1058	89	1088	96	1107	63	1124	47	1172	51
1267	54	1396	83	1426	82	1436	90	1466	89
1488	94	1564	72	1585	84	1625	53	1668	57
1907	67	3066	78						

COMPOUND : **Rotenone**

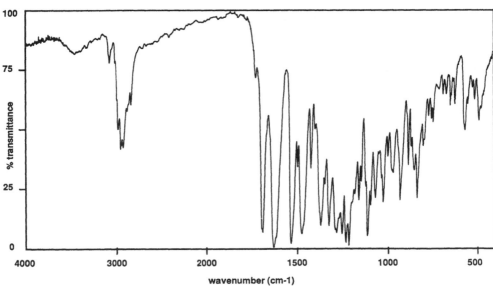

Chemical name	:	(2R,6aS,12aS)-1,2,6,6a,12,12a-hexahydro-2-isopropenyl-8,9-dimethoxychromeno [3,4b] furo [2,3-h] chromene-6-one		
Other names	:	Derris, Cube		
Type	:	insecticide, acaricide		
Brutoformula	:	C23H22O6	CAS nr	: 83-79-4
Molecular mass	:	394.4281	Exact mass	: 394.1416258
Instrument	:	Bruker IFS-85	Optical resolution	: 2 cm-1
Scans	:	32	Sampling technique	: KBr pellet

Band maxima with relative intensity :

476	45	501	36	551	49	607	38	634	39
655	34	674	34	729	46	738	44	754	43
783	56	812	78	829	66	846	56	864	64
911	79	951	67	982	60	1005	80	1049	78
1076	81	1091	94	1129	71	1141	79	1195	98
1212	97	1233	93	1262	93	1303	90	1351	90
1408	66	1455	94	1482	65	1514	98	1609	100
1672	93	2834	40	2912	58	2940	59	2970	50

COMPOUND : **2,3,5-Trimethacarb**

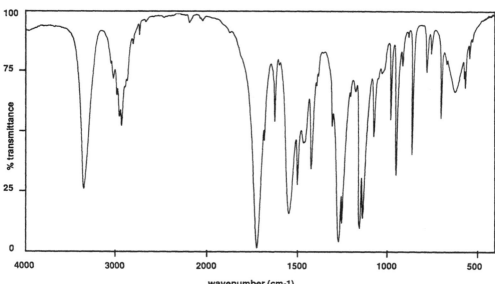

Chemical name   : 2,3,5-trimethylphenyl methylcarbamate

Other names   : -
Type   : insecticide, mulloscicide
Brutoformula   : C11H15NO2   CAS nr   : 3971-89-9
Molecular mass   : 193.247   Exact mass   : 193.110270
Instrument   : Bruker IFS-85   Optical resolution : 2 cm-1
Scans   : 32   Sampling technique : KBr pellet

Band maxima with relative intensity :

538	18	559	31	615	33	691	44	751	17
775	24	850	59	908	22	936	68	970	45
1063	52	1121	86	1136	91	1239	89	1255	97
1298	48	1412	66	1457	55	1488	73	1534	85
1622	46	1677	54	1711	100	2192	5	2736	10
2916	48	3017	29	3337	75				

406

Brutoformula Index

Alphabetical Index

–